Euclid—The Creation of Mathematics

Springer
New York
Berlin
Heidelberg
Barcelona
Hong Kong
London
Milan
Paris
Singapore
Tokyo

Benno Artmann

Euclid—The Creation of Mathematics

With 116 Illustrations

Drawings by the Author

 Springer

Benno Artmann
Pastor Sander Bogen 66
D-37083 Goettingen
Germany

Mathematics Subject Classifications (1991): 00A05, 01A20, 11A05, 51M05

Library of Congress Cataloging-in-Publication Data
Artmann, Benno.
    Euclid—the creation of mathematics / Benno Artmann.
      p.   cm.
    Includes bibliographical references and index.
    ISBN 0-387-98423-2 (alk. paper)
    1. Euclid, Elements. 2. Mathematics, Greek. I. Title.
QA31.A78 1999
510—dc21   98-31042

Printed on acid-free paper.

Production managed by A. Orrantia; manufacturing supervised by Nancy Wu.
Photocomposed copy prepared by Bartlett Press, Inc., Marietta, GA.
Printed and bound by Maple-Vail Book Manufacturing Group, York, PA.
Printed in the United States of America.

9 8 7 6 5 4 3 2 1

ISBN 0-387-98423-2 Springer-Verlag New York Berlin Heidelberg SPIN 10659097

Preface

This book is for all lovers of mathematics. It is an attempt to understand the nature of mathematics from the point of view of its most important early source.

Even if the material covered by Euclid may be considered elementary for the most part, the way in which he presents it has set the standard for more than two thousand years. Knowing Euclid's *Elements* may be of the same importance for a mathematician today as knowing Greek architecture is for an architect. Clearly, no contemporary architect will construct a Doric temple, let alone organize a construction site in the way the ancients did. But for the training of an architect's aesthetic judgment, a knowledge of the Greek heritage is indispensable. I agree with Peter Hilton when he says that genuine mathematics constitutes one of the finest expressions of the human spirit, and I may add that here as in so many other instances, we have learned that language of expression from the Greeks.

While presenting geometry and arithmetic Euclid teaches us essential features of mathematics in a much more general sense. He displays the axiomatic foundation of a mathematical theory and its conscious development towards the solution of a specific problem. We see how abstraction works and enforces the strictly deductive presentation of a theory. We learn what creative definitions are and how a conceptual grasp leads to the classification of the relevant objects. Euclid creates the famous algorithm that bears his name for the solution of specific problems in arithmetic, and he shows us how to master the infinite in its various manifestations.

One of the greatest powers of scientific thinking is the ability to uncover truths that are visible only "to the eyes of the mind," as Plato says, and to develop ways and means to handle them. This is what Euclid does in the case of irrational, or incommensurable, magnitudes. And finally, in the *Elements* we find so many pieces of *beautiful* mathematics that are easily accessible and can be studied in detail by anybody with a minimal training in mathematics.

Seeing such general phenomena of mathematical thinking that are as valid today as they were at the time of the ancient Greeks, we cannot but agree with the philosopher Immanuel Kant, who wrote in 1783 in the popular introduction to his philosophy under the heading "Is metaphysics possible after all?":

> There is no book at all in metaphysics such as we have in mathematics. If you want to know what mathematics is, just look at Euclid's *Elements*.
> (Prolegomena §4 in free translation)

To be sure, some fundamental concepts needed for today's mathematics are absent from Euclid's *Elements*, most notably algebraic formulae and the concepts of a function and the real numbers. This does not, however, affect us when we study, for instance, how the exclusion of certain means of construction forces a more rigorous treatment of a piece of mathematics or how generalizations can be very profound on one occasion and shallow on another one.

In order to substantiate these general claims, we have to look at Euclid's *Elements* in detail. Our comments are based on the translation from the Greek original by Th. L. Heath (1908/1926). We will quote many of Euclid's definitions and theorems, but cite proofs only for a few of the most important results like Pythagoras's theorem or the irrationality of $\sqrt{2}$. Many of Euclid's other proofs are reproduced in their main outlines and in a mildly modernized language to be more easily accessible to the reader. For any questions about specific details the interested reader is advised to have Heath's or any other translation into a modern language at hand. All recent translations are reliable, as my friends competent in classical philology tell me. If there are any problems in understanding a passage, they are almost always of a mathematical, and not a linguistic, nature and of minor importance. There are none of the ambiguities in a mathematical text as there are, for instance, in a novel or a philosophical treatise. However, in the *Elements* we still have a very distant historical text that needs interpretation. Here the present book takes a clear position: The Elements are read, interpreted, and commented upon from the point of view of modern mathematics.

I would like to share my joy and enthusiasm in studying Euclid with as many interested readers as possible. The *Elements* definitely deserve to be more widely known and read, and not just in the narrow circle of specialists in the history of mathematics. Nevertheless, I hope that the experts will find one or another aspect of my interpretation new and interesting.

Over the years, and during various stages of the preparation of this book, I have been helped and encouraged very generously by many people. Thanks to you all:

First and foremost, Maresia Artmann, then U. Artmann, L. Berggren, D. Fowler, C. Garner, J. Hainzl, R. Hersh, P. Hilton, K. H. Hofmann, W. Jonsson, V. Karasmanis, H. Knell, K. Lengnink, A. Mehl, Heike Müller, Ian Mueller, E. Neuenschwander, M. L. Niemann, S. Prediger, H. Puhlmann, L. Schäfer, M. Taisbak and his helping hand, S. Unguru, H.-J. Waschkies.

Benno Artmann
Goettingen, Germany

Contents

Notes to the Reader

A large part of Euclid's *Elements* treats a body of mathematics that today would be called elementary geometry and arithmetic. Even today, textbooks for a high-school course in geometry, for example the popular H. R. Jacobs "Geometry" [1987], follow, for large parts, Euclid's *Elements* rather closely. One can understand, for instance, what is said about similarity in Euclid's Book VI without reading much of the preceding books. While it is advisable to follow Euclid step by step, it is not absolutely necessary. A reader with a solid background in high-school geometry can enter the discussion at any particular place to follow a particular interest. Chapter I, containing a short description of the contents the thirteen books of the *Elements*, is included for the convenience of those readers who are interested in special subjects.

In our main text, the description and comments on each of Euclid's books will be followed by more general remarks about typical mathematical procedures, subjects of particular historical interest, connections to philosophy, and similar items.

The 13 books (= chapters) of the *Elements* are denoted by Roman numerals, the propositions in each Book by Arabic numerals. Hence Prop. II.14 is Proposition 14 of Book II. Definitions are indicated by Def. II.2.

Propositions and definitions are quoted from the translation of Euclid by T. L. Heath [1926] or, in a sometimes slightly modernized version, Mueller [1981]. As mentioned in the Preface, I have tried to explain the mathematical content of the *Elements* to readers with a general interest in mathematics and its history. For more specialized questions one should consult Heath's or any other modern translation. Aside from presenting the complete text, Heath discusses a great wealth of historical and other related details. He points out occasional little gaps in Euclid's arguments, and lists alternative proofs and similar items. He does not, however, speak about the general mathematical outlines of the text; he refrains from any judgment of the relative importance of the various propositions or theories and does not call to the attention of his readers general features of mathematics. All of these are of special interest for the present book.

Notes for the text are collected at the end of the book. There are very few numbered footnotes in the traditional sense. In the notes relevant books and papers are listed together with a few supplementary remarks. Whenever necessary, precise bibliographical information is given in the main text. As the first part of the notes a few books of general interest are listed.

Much of the vocabulary used in mathematics stems from Greek origin, words like "theorem" and "orthogonal." Aside from such familiar terms, there may be a few unfamiliar words. "Scholion" and "anthyphairesis" are the two of them that have to be explained most frequently. A *scholion* is a remark to the main text of the *Elements* that was written (mostly in antiquity) in the margin of old manuscripts. Scholia are somewhat like modern footnotes, but they are very valuable because they transmit information from antiquity. *Anthyphairesis* (or, equivalently, antaneiresis) is nothing but the Euclidean algorithm for general magnitudes, mostly for line segments.

Finally, the use of the name *Euclid* is ambiguous. On the one hand, it means the author of the *Elements* who lived about 300 B.C.E. On the other hand, "Euclid" stands for a collection of mostly unknown Greek mathematicians who lived between 500 and 300 B.C.E. and contributed most of the material contained in the *Elements* (and had themselves written "Elements" of some kind before Euclid). Used in that sense, a phrase like "Euclid proved ..." means just "a Greek mathematician proved...."

1

General Historical Remarks

The description "Ancient Greece" refers to the period roughly from 800 B.C.E. to 150 B.C.E., from Homer to the time when Rome established political hegemony over the Greek world. The first Olympic games took place in 776 B.C.E.; democracy was gradually introduced in the political life of the city-states from 600 B.C.E. onwards. The Greeks defended their freedom against the Persians in the "Persian wars" (500–480), after which the great classical period of Greece under the cultural leadership of Athens lasted until the Macedonian Kings Philip and Alexander the Great established monarchic rules around 330 and spread Greek culture over the whole ancient world in Hellenistic times, 300–50 B.C.E. Science and philosophy remained the domain of Greeks until the end of the Roman empire. Boetius, the "last Roman," was the first writer to translate mathematical texts from Greek into Latin, about 500 C.E. The Romans ran their Imperium without any mathematics.

Greek mathematics and philosophy begins with Thales of Miletus about 580 B.C.E., of whom very little is known. Certainly the Greeks learned some mathematical rules and procedures, and especially astronomy, from the old cultures in Egypt and Mesopotamia, but nothing has been found in these sources in the sense of mathematics as we know it today and that we encounter in Euclid's work.

1

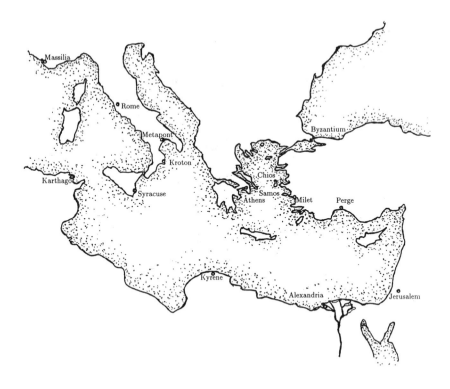

FIGURE 1.1

Mathematics was invented in Greek's classical period, beginning with Thales and Pythagoras about 530 B.C.E. and finding its final form with Euclid about 300 B.C.E. It stands on equal footing with the great Greek works in literature, sculpture, painting, architecture, historical and political writings, medicine, and philosophy.

2

CHAPTER

The Contents of the *Elements*

Traditionally, the *Elements* have been divided into three main parts:

1. Plane geometry, Books I–VI;

2. Arithmetic, Books VII–X;

3. Solid geometry, Books XI–XIII.

It will soon be obvious that Books V and X do not really fit into this division, but it is convenient to adhere to it in the following description of the individual books. The size of the books varies between ca. 2.5% of the whole for the smallest, Book II, and 25% for Book X. Each of the others is roughly 5–8% of the total.

2.1 Book I: Foundations of Plane Geometry

Book I starts with a set of definitions. Basic concepts such as point, line, angle are described in general terms and used to define various sorts of triangles, quadrangles, etc. The very last definition describes parallel lines in the plane as lines with no common point. After the definitions we find the so-called postulates, which are the axioms of geometry; the fifth and last of these is the famous parallel postulate. The "common notions" are axioms concerning magnitudes in general, e.g., "things equal to the same thing are equal to each other."

The theorems of Book I can be grouped into four sets, which will be discussed in greater detail below.

A. (I.1–26) Fundamental theorems and basic constructions in plane geometry such as the congruence theorems for triangles or the bisection of an angle; in part A no use is made of parallel lines.

B. (I.27–32) The theory of parallel lines, including the theorem that the sum of the interior angles of a triangle is equal to two right angles (I.32).

C. (I.33–45) The theory of parallelograms; transformation and comparison of areas of parallelograms and triangles.

D. (I.46–48) The theorem of Pythagoras.

2.2 Book II: The Geometry of Rectangles

Compared to Book I, the second book is very homogeneous. Most of the theorems of Book II can be interpreted as what we might call, in algebraic terms, variations on the theme of the binomial identity:

$$(a+b)^2 = a^2 + 2ab + b^2.$$

These results are always expressed in the geometric language of subdivisions of rectangles and of the areas of the various parts of the subdivisions. Theorems II.12 and 13 generalize the theorem of Pythagoras (I.47) to what we would call the law of cosines, and Proposition II.14 gives the solution of the important problem of constructing a square equal (in area) to a given rectilinear figure.

2.3   Book III: The Geometry of the Circle

Book III has neither the obvious subdivisions of Book I nor the homogeneous structure of Book II. After some definitions it presents the basic geometrical facts about circles, tangents, and circles in contact. The second half of Book III is in part concerned with what one could call the theory of quadrangles and circles, including Proposition III.21, which asserts the equality of all angles in the same segment of a circle.

2.4   Book IV: Regular Polygons in Circles

Book IV is by far the most homogeneous and tightly constructed book of the *Elements*. The following four problems are treated systematically:

(i) to inscribe a rectilinear figure in or (ii) circumscribe it about a given circle;

(iii) to inscribe a circle in or (iv) circumscribe it about a given rectilinear figure.

These problems are solved for triangles in general, squares, regular pentagons, hexagons and 15-gons.

2.5   Book V: The General Theory of Magnitudes in Proportion

Book V is the most abstract book in the *Elements* and is independent of the preceding books. Whereas the other books are concerned with either geometrical objects or numbers, this one treats "magnitudes," which include, according to Aristotle, numbers, lines, solids, and times. In VI.33 angles are treated as magnitudes, and plane areas figure as magnitudes in VI.1, XII.1, and XII.2 (areas of circles), as well as in many other places. This generality makes the theory of proportions applicable throughout mathematics, justifying Eratosthenes' statement that it is "the unifying bond of the mathematical sciences."

Various sources indicate that Eudoxus (about 400–350) created the theory in Book V.

Some of the theorems in Book V, such as

$$\text{V.16.} \quad a : b = c : d \Rightarrow a : c = b : d,$$

were almost certainly used (with different definitions and proofs) long before the time of Eudoxus.

2.6   Book VI: The Plane Geometry of Similar Figures

In its composition and general outlook Book VI is close to Book I. In fact, Books I, III, and VI represent the core of plane geometry, and their overall organization gives the impression of a standard treatment of geometry that has been reworked several times. The whole edifice of Book VI is based on Theorem VI.1 and its immediate consequence, VI.2, the fundamental theorem on the proportionality of line segments.

One of the main theorems connects lines and areas of similar triangles (and polygons) (VI 19,20): If triangles are similar with a similarity factor k for lines, then the factor for the corresponding areas is k^2.

The concluding section of Book VI deals with what Euclid calls "the application of areas," which in modern terms amounts to the geometric solution of quadratic problems. It can be translated into a modern treatment of quadratic equations. For this reason it has been called "geometric algebra" by some authors.

2.7 Book VII: Basic Arithmetic

With Book VII, Euclid starts afresh. Nothing from the preceding books is used, particularly none of the general theory of proportions of Book V. The definitions at the beginning of Book VII are intended to serve for all of VII–IX. Proportionality of numbers is defined in Def. 20 with no connection to Book V.

Euclidean arithmetic is founded on the Euclidean algorithm for determining whether two numbers are prime to one another (VII.1–4). In fact, the Euclidean algorithm gives the greatest common divisor (gcd) of any two numbers a, b. The next part of Book VII uses Def. 20 to establish the fundamental properties of proportions for numbers. To a great extent the more general theorems established in Book V are proved again for numbers. The mathematical core of Book VII is the theory of the gcd (VII.20–32), which has its counterpart in the theory of the least common multiple (lcm; VII.33–39).

The important subject of prime numbers in Euclid's arithmetic will be discussed in detail below.

2.8 Book VIII: Numbers in Continued Proportion

Whereas Book VII, like Book I, has a clear internal structure, Book VIII and the first part of IX are entangled like Book III. Numbers in continued proportion are the main focus of Book VIII.1–10. The second part (VIII.11–27) is concerned more with special types of numbers "in geometrical shape," such as squares and cubes. One important question in this context is how to characterize numbers a, b for which there exists a mean proportional x, i.e., an x such that

$$a : x = x : b.$$

2.9 Book IX: Numbers in Continued Proportion; the Theory of Even and Odd Numbers, Perfect Numbers

There is no break in subject matter between Books VIII and IX because IX.1 picks up where Book VIII ends. Euclid's other books have well-defined subjects, but in this case the division between VIII and IX looks artificial. Even more curious is that after IX.20 Euclid turns to a completely new subject, the theory of even and odd numbers ("the even and the odd," as Plato says), which has no connection with what precedes, but rests only on Definitions 6–10 of Book VII. This theory culminates in the construction of even perfect numbers (IX.36).

2.10 Book X: Incommensurable Line Segments

Book X is the most voluminous book of the *Elements*, occupying about one quarter of the whole work. In it, the Euclidean algorithm of Book VII is applied to general magnitudes in order to obtain the criterion for commensurability:

X.5. Commensurable magnitudes have to one another the ratio which a number has to a number.

X.6. If two magnitudes have to one another the ratio which a number has to a number, the magnitudes will be commensurable.

In X.9 Euclid states as an immediate consequence that the side of a square of area n is incommensurable with the side of a square of area 1 when n is not a square number. The bulk of the material of Book X, up to Proposition 115, consists in a careful study of various types of incommensurable lines and is beyond the scope of our intentions.

Historically, the discovery of incommensurable lines, or, as we would say today, irrational numbers, seems to have been of paramount importance. We will give an up-to-date description of what is known about this problem.

2.11   Book XI: Foundations of Solid Geometry

Book XI begins with a long list of definitions for Books XI–XIII. There are no postulates of the kind we find in Book I, so that there is no axiomatic foundation for Euclid's deductions at the beginning of Book XI. The general composition of Book XI is closely parallel to that of Book I. It has the following sections

A. (XI.1–19) Fundamentals of solid geometry (lines, planes, parallelism, and orthogonality).

B. (XI.20–23) Solid angles, their properties and construction.

C. (XI.24–37) Parallelepipedal solids.

2.12   Book XII: Areas and Volumes; Eudoxus's Method of Exhaustion

Some infinitesimal methods are needed to determine the area of a circle in relation to a square, or the volume of a pyramid. The method of exhaustion, which Euclid employs, is said to have been first applied rigorously by Eudoxus, to whom most of the contents of Book XII are attributed. The method of proof is quite different from—and much more intricate than—anything in the preceding geometrical books, with the exception of Book V.

2.13 Book XIII: The Platonic Solids

The first part of Book XIII, that is, XIII.1–12, consists of various
planimetric propositions. Some of them are evidently lemmas for
the subsequent theory of the regular polyhedra, and some others
are concerned more generally with the division of a straight line
in extreme and mean ratio (much later called the golden section).
Because division in extreme and mean ratio is indispensable for the
constructions of the icosahedron and the dodecahedron, and these
are the subject of the last part of Book XII, we may regard the whole
first part of Book XIII as preparatory for the second.

Each of the regular solids is treated in a separate two-part
theorem:

(i) To construct the solid and to comprehend it in a given sphere.

(ii) To compare the diameter of the sphere with the side of the
polyhedron, in the sense of the classification in Book X.

Euclid's treatment of the regular polyhedra is especially important for the history of mathematics because it contains the first
example of a major classification theorem. Moreover, the regular
polyhedra have always been of interest for mathematicians. They
also play a major role in Plato's philosophy. These topics will be
discussed in an appendix to Book XIII.

3
CHAPTER

The Origin of Mathematics 1

The Testimony of Eudemus

....

Our most important source about the history of Greek mathematics before Euclid originates from Eudemus of Rhodes, a student of Aristotle, who lived from about 350 to 300 B.C.E. He wrote a book on the history of mathematics, which has, however, been lost except for a few passages quoted by other authors. The following one was preserved in Proclus's *Commentary on Euclid*. Proclus (410–485 C.E.) is writing about the origin and development of geometry, and the Eudemus passage starts with the second paragraph:

> Limiting our investigation to the origin of the arts and sciences in the present age, we say, as have most writers of history, that geometry was first discovered among the Egyptians and originated in the remeasuring of their lands. This was necessary for them because the Nile overflows and obliterates the boundary lines between their properties. It is not surprising that the dis-

covery of this and the other sciences had its origin in necessity, since everything in the world of generation proceeds from imperfection to perfection. Thus they would naturally pass from sense-perception to calculation and from calculation to reason. Just as among the Phoenicians the necessities of trade and exchange gave the impetus to the accurate study of number, so also among the Egyptians the invention of geometry came about from the cause mentioned.

Thales, who had travelled to Egypt, was the first to introduce this science into Greece. He made many discoveries himself and taught the principles for many others to his successors, attacking some problems in a general way and others more empirically. Next after him Mamercus, brother of the poet Stesichorus, is remembered as having applied himself to the study of geometry; and Hippias of Elis records that he acquired a reputation in it. Following upon these men, Pythagoras transformed mathematical philosophy into a scheme of liberal education, surveying its principles from the highest downwards and investigating its theorems in an immaterial and intellectual manner. He it was who discovered the doctrine of proportionals and the structure of the cosmic figures. After him Anaxagoras of Clazomenae applied himself to many questions in geometry, and so did Oenopides of Chios, who was a little younger than Anaxagoras. Both these men are mentioned by Plato in the Erastae as having got a reputation in mathematics. Following them Hippocrates of Chios, who invented the method of squaring lunules, and Theodorus of Cyrene became eminent in geometry. For Hippocrates wrote a book on elements, the first of whom we have any record who did so.

Plato, who appeared after them, greatly advanced mathematics in general and geometry in particular because of his zeal for these studies. It is well known that his writings are thickly sprinkled with mathematical terms and that he everywhere tries to arouse admiration for mathematics among students of philosophy. At this time also lived Leodamas of Thasos, Archytas of Tarentum, and Theaetetus of Athens, by whom the theorems were increased in number and brought into a more scientific arrangement. Younger than Leodamas were Neoclides and his pupil Leon, who added many discoveries to those of their predecessors, so that Leon was able to compile a book of elements more carefully designed to take account of the number of propositions that had been proved and of their utility. He also discovered diorismi, whose purpose is to determine when a problem under investigation is capable of solution and when it is not. Eudoxus of Cnidus, a little later than Leon and a member of Plato's group, was the first to increase the number of the so-called general theorems; to the three proportionals already known he added three more and multiplied the number of propositions concerning the "section" which had their origin in Plato, employing the method of analysis for their solution. Amyclas of Heracleia, one of Plato's followers, Menaechmus, a student of Eudoxus who also was associated with Plato, and his brother Dinostratus made the whole of geometry still more perfect. Theudius of Magnesia had a reputation for excellence in mathematics as in the rest of philosophy, for he produced an admirable arrangement of the elements and made many partial theorems more general. There was also Athenaeus of Cyzicus, who lived about this time and became eminent in other branches of mathematics and most of all in geometry. These men lived together in the Academy, making their inquiries in common. Hermotimus of Colophon pursued further the investigations already begun by Eudoxus and Theaetetus, discovered many propositions in the *Elements*, and wrote some things about locus-theorems. Philippus of Mende, a pupil whom Plato had encouraged to study mathematics, also carried on his investigations according to Plato's instructions and set himself to study all the problems that he thought would contribute to Plato's philosophy.

All those who have written histories bring to this point their account of the development of this science. Not long after these men came Euclid, who brought together the *Elements*, systematizing many of the theorems of Eudoxus, perfecting many of those of Theaetetus, and putting in irrefutably demonstrable form propositions that had been rather loosely established by his predecessors. He lived in the time of Ptolemy the First ... (Proclus–Morrow pp. 51–56)

Let us underline some of the specific characteristics of this quotation. At the beginning, the text acknowledges the foreign origin of mathematics: The Greeks learned geometry ("land measurement") from the Egyptians and arithmetic from the Phoenicians. The alleged Egyptian origin of geometry strikes us as implausible: We don't know any Egyptian text that would substantiate this claim, but it does occur several times in Greek sources. Thales, one of the proverbial wise men of Greece, is characterized as the intermediator between the "barbarians" and the Greeks. Pythagoras gets the credit for the decisive transformation of mathematics into an abstract science and, moreover, into a subject of liberal education, but there are doubts about the origin of the passage about the doctrine of proportionals and the cosmic figures, i.e., the regular solids. This may be an interpolation by Proclus.

The very strong relations between mathematics and philosophy are stressed in the remarks about Plato. He is described as a sort of research director, who instructs the mathematicians what to do. Both Theaetetus and Leon are praised for bringing theorems "into a more scientific arangement." Theudius of Magnesia became famous because he "made many partial theorems more general."

At the beginning of the last paragraph there is the hint that the passage is by Eudemus, though he is not mentioned by name, and the remarks about Euclid seem to be by Proclus himself. Some specific details of this report are disputed, especially in its tendency to overestimate the role of Pythagoras and Plato. On the other hand, we would know virtually nothing about many of the mathematicians mentioned had Proclus not preserved this passage for us.

For the moment let us note that Eudemus mentions some writers of "Elements" earlier than Euclid, the first of whom is Hippocrates of Chios about 440–430 B.C.E. (He is not to be confused with the medical man Hippocrates of Kos.) Leon compiled the textbook for Plato's Academy (370 B.C.E.?), and Theudius "produced an admirable arrangement of the elements." In due course we will discuss the contribution of various mathematicians to Euclid's *Elements*.

What Does the Word "MATHEMATICS" Mean?

This whole book is about mathematics, but here we are looking only at the etymological side of the question. The Greek word *mathema* ($\mu\alpha\theta\eta\mu\alpha$) originally means "that which is learned, learning, science" and was first used in this sense by Plato and, probably, the Pythagoreans. The associated verb is the Greek *manthanein*, to learn.

The word is derived from the Indo-European root mendh- "to have one's mind aroused, apply oneself to." Here are some related words from other languages:

English: *mind*

German: *munter* (awake, lively, merry, vigorous)

Middle High German: *Minne* (love)

Gothic: *munda* (to aim)

Old Slavic: *modru* (wise, sage)

Sanskrit: *man* (to think)

Latin: *mens* (mind)

Greek: *mantis* (a seer),

and possibly even the Greek *muse* and *Prometheus*. We mathematicians find ourselves among a host of very sympathetic counterparts.

TIME TABLE (all dates are B.C.E.)		
General history, related to mathematics		Mathematics
900–600 Geometric period of Greek art		
Money is invented (first coins)	600	Thales of Miletus ($\approx$ 580)
		Pythagoras ($\approx$ 570–490)
Persian Wars ($\approx$ 500–480)	500	
$\approx$ 460 Temple of Zeus in Olympia, Proportions 2 : 1		Pythagoreans in southern Italy
$\approx$ 450–430 Pericles, "high classics"	450	
$\approx$ 440 Parthenon temple in Athens, proportions 9 : 4 = length : breadth = breadth : height		Hippocrates of Chios writes first *Elements* $\approx$ 430
Socrates $\approx$ 470–399		Theodorus of Cyrene $\approx$ 460–390
Plato 428–348	400	Theaetetus $\approx$ 415–370 Leon writes new *Elements* in Plato's Academy
Aristotle 384–322		
		Eudoxus $\approx$ 410–355
Alexander the Great 356–323	350	Various other writers of mathematical treatises, e.g., Menaechmus: *Conics*
	300	Euclid: *Elements*
Alexandria is the cultural center of the Hellenistic world 300–50	250	Apollonius of Perga, Archimedes of Syracuse

4

Euclid Book I

Basic Geometry

....

4.1   The Overall Composition of Book I

Definitions 1–23	Basic concepts are described or defined
Postulates 1–5 Common Notions 1–5	Postulates and Common Notions are the axioms of plane geometry
1–26	A: Foundations of plane geometry without using parallels
27–32	B: The theory of parallel lines, angles in a triangle
33–45	C: The theory of parallelograms and their areas
46–48	D: The theorem of Pythagoras

4.2   Definitions and Axioms

Euclid created the model of a mathematical text: Start with explicitly formulated definitions and axioms, then proceed with theorems and proofs. Unlike modern authors, who do not pretend to know what a set is, Euclid wants to say what he is talking about, or to give some sort of description of the objects of geometry. He does this in the first group of definitions, 1–9.

Definitions
Def. 1. *A point is that which has no part.*
Def. 2. *A line is breadthless length.*
Def. 3. *The extremities of a line are points.*

. . .

Def. 8. *A plane angle is the inclination to one another of two lines in a plane which meet one another and do not lie in a straight line.*
Def. 9. *And when the lines containing the angle are straight, the angle is called rectilinear.*

It has often been observed that Euclid makes no use of these definitions in his subsequent proofs. They are explications that should clarify the significance of a term to the reader but play no formal rule in deductions. In Def. 8, the lines forming an angle may be curved. In Book III Euclid occasionally uses angles between circles and straight lines, but in our discussion of Proposition I.5 we will find indications for a greater popularity of angles between curved lines in pre-Euclidean times.

Most of the following definitions are abbreviations in the modern manner, for instance:

Definitions
Def. 19. ... *trilateral figures are those contained by three straight lines* ...
Def. 20. *Of trilateral figures, an equilateral triangle is that which has its three sides equal, an isosceles triangle which has two of its sides alone equal, and a scalene triangle that which has its three sides unequal.*

In a modern formal sense, an equilateral triangle is isosceles as well, but not so for Euclid. Similarly, in Def. 22, a rectangle (called "oblong" there) is not a square. This agrees with a more colloquial modern use: If a rectangle is spoken of, in most cases this means "not a square," because otherwise one could be more specific. Obviously, from a logical point of view, it is much more convenient to include the squares with the rectangles.

After the definitions, Euclid proceeds to state his famous postulates. Modern axioms of geometry resemble these postulates rather closely.

Postulates
1. *Let it be postulated to draw a straight line from any point to any point, and*
2. *to produce a limited straight line in a straight line,*
3. *to describe a circle with any center and distance,*
4. *that all right angles are equal to each other.*
5. *[The parallel postulate will be discussed below]*

Today Postulates 1 and 2 would be expressed in a way like "given any two distinct points, there is a unique line passing through them." Euclid's emphasis is more on construction than on "existence," more a difference in style than in substance.

The geometric postulates are followed by what Euclid calls "common notions." These are axioms about the behavior of general magnitudes, not only geometric objects.

1. *Things equal to the same thing are also equal to one another.*
2. *If equals are added to equals the wholes are equal.*
3. *If equals are subtracted from equals the remainders are equal.*
4. *Things which coincide with one another are equal to one another.*
5. *The whole is greater than the part.*

Many authors have noted the incompleteness of Euclid's axioms in comparison to modern foundations of geometry. The most obvious point is the absence of any thought of the ordering of points on a line or the concept of "betweenness." Euclid uses all assertions about ordering on an intuitive basis. These objections concern relatively minor points and do not in any way diminish Euclid's basic achievement: In mathematics, one has to start from explicitly stated first principles and deduce all following assertions from these principles.

Historically, the idea of stating axioms seems to be rather new in Euclid's time when compared to the definitions. In a very thorough and penetrating investigation, Mueller [1991] examines the starting points of mathematical theories as preserved in the writings of Plato, Aristotle, and Euclid. Mueller summarizes on p. 63:

> However, if we look at the *Elements*, although we find at the beginning of book 1 definitions, postulates, and common notions . . . at the beginning of the remaining books we find only definitions. I believe there are two related inferences we can draw from this: (1) Euclid did not believe that proportion theory, number theory, or solid geometry required its own postulates; (2) at the end of the fourth century there were no accepted presentations of these theories which included postulates, and probably no such presentations at all, presumably because no mathematician recognized the need for them. A further inference I draw is that the idea of such presentations of any mathematical theory was relatively new in Euclid's time, i.e., did not precede Plato's maturity. I believe the evidence suggests that Euclid himself is responsible for the postulates, but for the moment I will only say that, even if they are thought to predate, say, Plato's *Republic*, they should still be seen as the exception rather than the rule by Euclid's time.
>
> The rule in the *Elements* and, I am suggesting, earlier in the history of Greek mathematics is a theory, the only explicit starting points of which are definitions. These definitions are, for the most part, either explications, which perhaps clarify the significance of a term to the reader but play no formal role in subsequent arguments, or abbreviations in the modern manner.

In spite of this rather diverse historical picture, Euclid's axioms have been of utmost importance for the development of mathematics because, as it was said at the beginning, with them he created the model of a mathematical theory.

4.3 Book I, Part A: Foundations

The essential contents of Part A of Book I are first the basic congruence theorems and elementary constructions such as bisecting angles and segments, and second some propositions about "greater" relations of angles and sides of triangles, based on I.16 and culminating with the triangle inequality I.20.

The very first propositions show how to construct an equilateral triangle and how to copy segments without moving them. The delicate constructions in I.2, 3 are based directly on the Postulates 1, 2, and 3. Proposition I.4 is the first substantial theorem, the congruence theorem "side–angle–side," for short, SAS. Euclid states it like this:

Prop. I.4.
If two triangles have the two sides equal to two sides respectively, and have the angles contained by the equal straight lines equal, they will also have the base equal to the base, the triangle will be equal to the triangle, and the remaining angles will be equal to the remaining angles respectively, namely those which the equal sides subtend.

For the proof see Fig. 4.1:

Let ABC, DEF be two triangles having the two sides AB, AC equal to the two sides DE, DF respectively, namely AB to DE and AC to DF, and the angle BAC equal to the angle EDF.

I say that the base BC is also equal to the base EF, the triangle ABC will be equal to the triangle DEF, and the remaining angles will be equal to the remaining angles respectively, namely those which the equal side subtend, that is, the angle ABC to the angle DEF, and the angle ACB to the angle DFE.

Before looking into the proof, we observe some peculiarities of Euclid's style. He always states his theorems in two ways: at first in general words, and then a second time in a more specific way indicating points, lines, angles, and so on by various letters.[1] Very often

---

[1] This is very much like today's usage: Theorem: A continuous real function maps closed intervals onto closed intervals. Let $[a, b]$ be a closed interval and $f : [a, b] \to \mathbb{R}$ be continuous ...

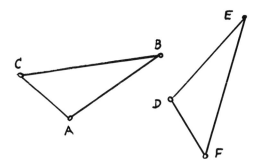

FIGURE 4.1

the theorem is accompanied by a suitable diagram. One particular phrase needs explanation: *"The triangle ABC will be equal to the triangle DEF."* This is clarified by later use of the same expression: It simply means "the triangles have equal areas." Euclid uses the word "area"(or its Greek equivalent) only occasionally.[2]

The Proof of Prop. I.4

The method of proof of I.4 stands in strong contrast to the meticulous constructions in I.1–3. Euclid just takes the triangle *ABC* and superimposes it on triangle *DEF* in a such way that *A* is placed on *D*, *B* on *E*, and *C* on *F*. From this he easily derives his assertions.

On the one hand, this method of superposition clearly has no basis in Euclid's axioms, but on the other hand, practically nothing can be done in elementary geometry without the congruence theorems. (For the congruence theorem side–side–side, SSS, in I.8 he uses the same method.) In fact, what we see here is another axiom. Modern axiomatic studies by Hilbert and others have shown that there is no way to resolve this dilemma: Either SAS has to be used as an axiom or one has to use superposition in a modern version by postulating the existence of certain rigid motions of the plane. (For more details, see Hartshorne Chapter 1.3.)

---

[2]The Greeks knew perfectly well how to measure their properties, and they knew that Pharaoh's taxation office measured the fields of the Egyptian peasants to their disadvantage. In mathematics they avoid the concept of "area," using instead phrases like the one above, e.g., "this rectangle is equal to that rectangle" and similarly.

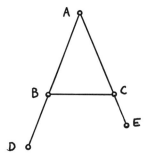

FIGURE 4.2

In the next pair I.5/6 of propositions, Euclid proves a fundamental lemma about isosceles triangles that is used frequently in Books I–VI. In the statement and proof of I.5 we ignore Euclid's assertion about outer angles. I.6 is the converse of I.5.

Prop. I.5

In isosceles triangles the angles at the base are equal to one another.

Let ABC be an isosceles triangle having the side AB equal to the side AC; and let the straight lines BD, CE be produced further in a straight line with AB, AC.

I say that the angle ABC is equal to the angle ACB and the angle CBD to the angle BCE [Fig. 4.2].

Prop. I.6.

If in a triangle two angles are equal to one another, the sides which subtend the equal angles will also be equal to one another.

For the proof of I.5, Euclid first constructs two auxiliary triangles *BFC* and *CGB* [Fig. 4.3]:

Let a point *F* be taken at random on *BD*; from *AE* the greater let *AG* be cut off equal to *AF* the less; and let the straight lines *FC*, *GB* be joined.

In the next two steps he first shows the congruence of the triangles $\triangle AFC$ and $\triangle AGB$ by using SAS, and then again by SAS the congruence $\triangle BFC \cong \triangle CGB$:

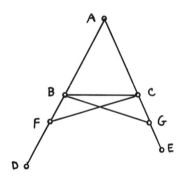

FIGURE 4.3

(1) We have $\angle FAC = \angle GAB$, and $AF = AG$ by construction, and $AC = AB$, hence $\triangle AFC \cong \triangle AGB$, and especially $BG = CF$ and $\angle BFC = \angle CGB$.

(2) From the construction we get $BF = CG$; moreover, we know from the outset $BC = CB$, and from (1) we have $\angle FBC = \angle CGB$; hence $\triangle CGB \cong \triangle BFC$ by SAS.

Now Euclid concludes:

Therefore the angle FBC is equal to the angle GCB, and the angle BCF to the angle CBG. Accordingly, since the whole angle ABG was proved equal to the angle ACF, and in these the angle CBG is equal to the angle BCF, the remaining angle ABC is equal to the remaining angle ACB; and they are at the base of the triangle ABC.   Q.E.D.

All the steps of this proof are justified by Euclid's axiomatic base. (For details, see the analysis in Hartshorne Ch. 2.10.)

We will direct our attention to another question, which has frequently baffled students of Euclid: How can anybody understand the introduction of his auxiliary points, lines, and triangles at the beginning of his proof?

In this particular instance, I.5, we are in the lucky position of having a historical predecessor of Euclid's proof that explains the initial construction. It comes from Aristotle's *Prior Analytics*, we quote it from Heath's commentary on I.5. It makes use of mixed angles between circular arcs and straight lines in the following way: (a) the angles of semicircles (called AC and BD), that is, between a diameter and the circumference, are equal, and (b) the two angles in a segment, that is, between a chord and the circumference, are equal. Aristotle uses the proof in his discussion of some logical points. (See Fig. 4.4.)

> For let A, B be drawn [i.e. joined] to the center.
> If then, we assumed (1) that the angle AC is equal to the angle BD without asserting generally that the angles of semicircles are equal, and again (2) that the angle C is equal to the angle D without making the further assumption that the two angles of all segments are equal, and if we then inferred, lastly, that, since the whole angles are equal, and equal angles are subtracted from them, the angles which remain, namely E, F are equal. We should commit a petitio principii, unless we assumed [generally] that, when equals are subtracted from equals, the remainders are equal. (*Prior Analytics* 41 b 13–22)

First observe a basic similarity in Aristotle's and Euclid's proofs: We have two equal big angles, from which two smaller equal angles are subtracted, resulting in the desired equality of the base angles.

In Euclid's time mixed angles were no longer acceptable; he does not use them save on a few minor occasions in Book III. A transition from Aristotle's proof to one without mixed angles can be explained in a plausible way. (We use Euclid's notation; see Fig. 4.5)

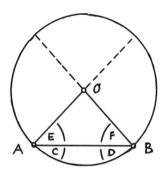

FIGURE 4.4

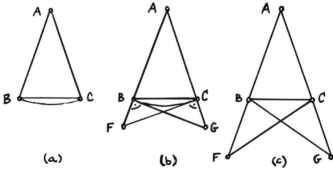

FIGURE 4.5

First replace the "angle of the semicircle" by the right angle between the radius AB and the tangent BG, similarly for AC and CF.
The congruence theorem ASA (With α, side AC = side AB, and the
right angles) would be needed in order to show $\triangle ABG \cong \triangle ACF$.
Hence the "big" right angles $\angle ABG$ and $\angle ACF$ are equal, and the
result would follow as in Euclid's proof. Note that the symmetri-
cally situated small triangles $\triangle BFC$ and $\triangle CBG$ replace the intuitively
symmetric segment.

However, Euclid cannot use tangents and ASA at this stage of
Book I, so he disposes with the right angles between radii and tangents and cleverly provides himself with the equal sides AF and AG
in a direct way so that he can use SAS instead of ASA. (E and D are
merely auxiliary points for the prolongation of the sides.)

Here, I think, we have found a natural explanation for Euclid's
construction, albeit a hypothetical one. Observe that Fig. 4.5 (b) appears in Book III.17 where Euclid constructs the tangent to a circle
and, in a less obvious way, in I.2 as well.

Aristotle lived 384–322 and was a member of Plato's Academy
367–348 when Plato died. It seems very likely that he got his mathematical education in the Academy, and so it is possible that in
looking at his proof we see a small fragment of Leon's "Elements,"
the textbook of Plato's Academy.

Propositions 7–15. In Propositions 7 and 8 Euclid proves the
congruence theorem side–side–side (SSS), using the method of superposition for the second time. Propositions 9–15 are devoted

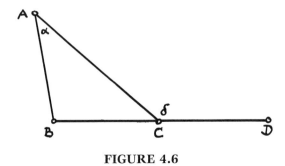

FIGURE 4.6

to the common auxiliary constructions and initial propositions of plane geometry: bisecting angles and segments, constructing perpendiculars, supplementary and vertical angles.

Prop. I.16.

If one of the sides of any triangle is produced, the exterior angle is greater than each of the interior and opposite angles.

Claim. angle α < angle δ (Fig. 4.6).

Construction. Bisect AC at E, draw BE and extend it to F such that $BE = EF$, join C and F, let $\alpha' =$ angle ECF (Fig. 4.7).

Proof

(i) Triangle ABE is congruent to triangle CFE by the congruence theorem SAS. Hence $\alpha = \alpha'$.

(ii) But α' is a part of δ. Hence $\alpha = \alpha' < \delta$ by common notion 5, Q.E.D.

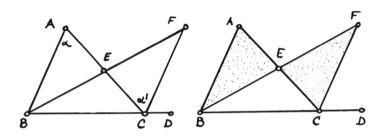

FIGURE 4.7

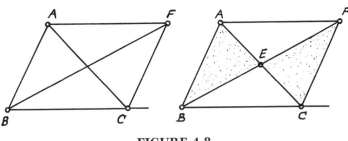

FIGURE 4.8

If Euclid had the theory of parallels at his disposal right here, the claim of I.16 would be a trivial consequence of I.32 about the sum of the angles in a triangle. He explicitly states in I.32 that the exterior angle is the sum of the two interior and opposite angles. Hence we see a conscious composition at work. Before discussing this, we will try to understand the genesis of the proof of I.16 with the help of parallels.

What can be said about this proof? It is ingenious, and one can see how its author hit upon his idea: Just add the line AF to the figure (Fig. 4.8).

All of a sudden, we see a parallelogram $ABCF$ "behind" the proof of I.16. At this stage, we may use parallels and have $\alpha = \alpha'$ because of alternate angles; AC is a transversal of the two parallel lines AF and BC. Furthermore, E will be the intersection of the diagonals of this parallelogram.

However, and this is the essential idea, in order to prove I.16 it is possible to avoid parallels and use the congruence theorem 1, 4 instead.

Further evidence of the mathematical competence of the author of I.16 is his ability to connect I.16 with its consequences, the important theorems I.20, the triangle inequality, and I.27, the existence of parallels. (After all, deductive structures are what mathematics is all about.)

On the other hand, there is a weak spot in the proof. The assertion "α' is a part of δ" has no base in Euclid's axioms. It is just read off from the diagram. This has often been observed: Compare, for instance, Heath's commentary. (Heath confuses the "Riemann hy-

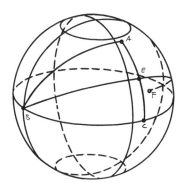

FIGURE 4.9

pothesis" with Riemannian non-Euclidean geometry, but otherwise he is mathematically correct.) Proposition I.16 is not true in the so-called elliptical (spherical) geometries, which satisfy all of Euclid's axioms except the parallel postulate. A counterexample to I.16 is easily drawn on the sphere. In Fig. 4.9, supplied by E. Hartmann, the point F will be in the southern hemisphere; hence $\alpha' > \delta$.

It should, however, be clear that any Greek mathematician would reply to this objection that he was dealing with plane, not spherical, geometry. Certainly a man like Menelaus of Alexandria (about 100 C.E.), who wrote about spherical geometry, knew the phenomenon. It seems that nobody noted the error before the end of the nineteenth century, when non-Euclidean geometries and order-relations in geometry came to the attention of mathematicians. The likely reason for Euclid's neglect of questions about the ordering of points on a line (or betweenness) may be that he regarded it as a part of logic—or just took it for granted. In fact, I.16 remains valid in the second class of non-Euclidean geometries, the so-called hyperbolic geometries, which can be defined over ordered fields. The reader interested in more details about order relations in geometry should consult Hartshorne, Chapters 1.3 and 3.15.

Propositions I.17–20. Proposition I.17 is a direct consequence of I.16. It is again a weak variant of I.32 about the sum of the angles in a triangle:

Prop. I.17.
In any triangle two angles taken together in any manner are less than two right angles.

Proposition I.18 says that in any triangle the greater side subtends the greater angle, and I.19 is its converse. These propositions lead to

Prop. I.20.
In any triangle two sides taken together in any manner are greater than the remaining one.

This is the famous triangle inequality. Proclus comments on this:

> The Epicureans are want to ridicule this theorem, say it is evident even to an ass and needs no proof ... they make [this] out from the observation that, if hay is placed at one extremity of the sides, an ass in quest of provender will make his way along the one side and not by way of the two others. (Proclus–Morrow p. 251)

(The Epicureans of today might as well add that one could see the proof on every campus where people completely ignorant of mathematics traverse the lawn in the manner of the ass.) Proclus replies rightly that a mere perception of the truth of a theorem is different from a scientific proof of it, which moreover gives reason why it is true. In the case of Euclid's geometry, the triangle inequality can indeed be derived from the other (equally plausible) axioms. On the other hand, the Epicureans win in the modern theory of metric spaces, where the triangle inequality is the fundamental axiom of the whole edifice.

Propositions I.21–26. Three of the remaining propositions of part A are about "greater" relations for sides and angles of triangles (21, 24, 25). Prop. 22 gives the construction of a triangle from its sides, provided that the triangle inequality is valid. Using this, Euclid shows in Prop. 23 how to copy an angle. The combined congruence theorems *ASA* and *AAS* are tagged on in I.26 as a sort of loose end.

4.4   Book I, Part B: The Theory of Parallels

Euclid defines in *Def.* I.23:

Parallel straight lines are straight lines which, being in the same plane and being produced indefinitely in both directions, do not meet one another in either direction.

For short, parallels in a plane are nonintersecting straight lines. This has remained unchanged in the modern theory of incidence geometry (cf. Hartshorne, Ch. 2.6). In so-called affine planes, parallel lines are those that have no point in common. The modern parallel axiom in an affine plane is this:

Given a line g and a point P not on g, there exists one and only one line h passing through P that does not meet g.

This axiom really has two parts:
(1) The parallel h to g through P exists.
(2) It is unique ("only one line").
(Part (2) is sometimes called "Playfair's axiom.")

Euclid's geometry is richer than the theory of affine planes: He has the congruence axioms and—implicitly—the conditions of ordering and betweenness. Via ordering he got I.16 and from this he derives the "existence" part for parallels in I.27. For uniqueness, he has to introduce a special axiom, the famous parallel postulate. He uses it in I.29 in order to prove a property of parallels that immediately provides uniquenes (without saying so). We quote Postulate 5 below, but first I.27 together with its proof.

Prop. I.27
If a straight line falling on two straight lines makes the alternate angles equal to one another, the straight lines will be parallel to one another.

For let the straight line *EF* falling on the two straight lines *AB*, *CD* make the alternate angles *AEF*, *EFD* equal to one another [Fig. 4.10]; I say that *AB* is parallel to *CD*.

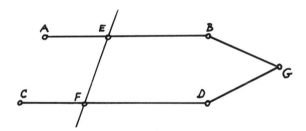

FIGURE 4.10

For, if not, *AB*, *CD* when produced will meet either in the direction of *B*, *D* or towards *A*, *C*.
Let them be produced and meet, in the direction of *B*, *D* at *G*.
Then, in the triangle *GEF*, the exterior angle *AEF* is equal to the interior and opposite angle *EFG*:
which is impossible.   [I.16]
Therefore *AB*, *CD* when produced will not meet in the direction of *B*, *D*.
Similarly it can be proved that neither will they meet towards *A*, *C*.
But straight lines which do not meet in either direction are parallel;
[Def. 23]
therefore *AB* is parallel to *CD*.

Before going on we introduce some convenient notation:

$$g \parallel h \qquad \text{for parallel lines} \qquad g \text{ and } h$$

and

$$2R \qquad \text{for two right angles (or } 180°\text{).}$$

Proposition 28 is a useful variant of 27. It says, with the notation taken from Fig. 4.11,

$$\alpha = \gamma \Longrightarrow g \parallel h,$$
$$\beta + \gamma = 2R \Longrightarrow g \parallel h.$$

For the proof of Prop. 29, we need the parallel postulate.

Postulate 5. Let it be postulated:

That, if a straight line falling on two straight lines makes the interior angles on the same side less than two right angles, the two straight lines,

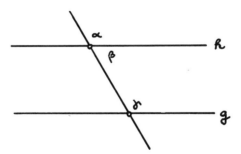

FIGURE 4.11

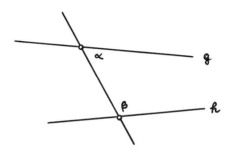

FIGURE 4.12

if produced indefinitely, meet on that side on which are the angles less than the two right angles. [Fig. 4.12]

Prop. I.29.
A straight line falling on parallel straight lines makes the alternate angles equal to one another, the exterior angle equal to the interior and opposite angle, and the interior angles on the same side equal to two right angles.

We will abbreviate the proof by using the notation from Fig. 4.13(b) and dealing with the main case of alternating angles α, β only.

The claim is:

$$\text{If} \quad g \parallel h, \quad \text{then} \quad \alpha = \beta.$$

This is the converse of I.27.

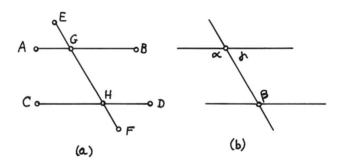

FIGURE 4.13

The proof is by contradiction, or rather by proving the logically equivalent statement

$$\text{If not} \quad \alpha = \beta, \quad \text{then not} \quad g \parallel h.$$

If not $\alpha = \beta$, then one of them is greater, say $\alpha > \beta$:

$$\beta < \alpha \Longrightarrow \beta + \gamma < \alpha + \gamma.$$

But

$$\alpha + \gamma = 2R,$$

whence

$$\beta + \gamma < 2R = \alpha + \gamma.$$

Now, Postulate 5 says that g and h have to meet, that is,

$$\beta + \gamma < 2R, \quad \text{and Post. 5} \quad \Longrightarrow \text{not } g \parallel h.$$

(If you want to see a contradiction like the one in Euclid's proof, continue: but $g \parallel h$ by hypothesis. . . .)

Propositions 27 and 29 together give us the fundamental property of parallels (*PP*), again with the notation of figure 13(b):

$$(PP) \qquad g \parallel h \Longleftrightarrow \alpha = \beta \qquad \text{(for alternating angles)}.$$

$g \parallel h$ has the intuitive meaning "g and h have no common point," a property that has to be checked from here to infinity. On the other side, $\alpha = \beta$ can be verified locally and is a most useful practical device that connects the intuitive notion with the other concepts of congruence geometry. This is a very common feature of mathematics: Define a concept by an intuitive meaning (if possible), and then prove that this is equivalent to a technically useful other statement. Another striking example of this is the definition of a tangent to a circle in Book III. In modern mathematics, unfortunately, the technical devices are often put in the foreground at the expense of the intuitive meaning.

Proposition 30 shows the transitivity of parallelism, and Proposition 31 exploits (PP) for the construction of parallels by alternate angles.

Prop. I.32.

In any triangle, if one of the sides is produced, the exterior angle is equal to the two interior and opposite angles, and the three interior angles of the triangle are equal to two right angles.

Again, we will take the convenient notation from Fig. 4.14 for Euclid's proof. Prolong the line BC to the point D. Let EC be parallel to AB. Then, by Prop. 29, we have

$$\alpha = \alpha' \qquad \text{and} \qquad \beta = \beta'.$$

Hence the exterior angle $\alpha' + \beta'$ is equal to $\alpha + \beta$, and because of $\gamma + \alpha' + \beta' = 2R$, we have

$$\alpha + \beta + \gamma = 2R.$$

Euclid emphasizes the exterior angle $\alpha' + \beta'$ because he will use it on several later occasions.

The sum of the angles of a triangle is the most important and fundamental invariant in elementary geometry. No matter what the

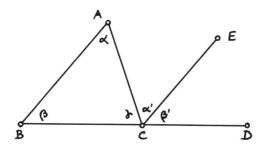

FIGURE 4.14

shape of the triangle is, its angles will invariably add up to two right angles (or 180 degrees, or π). This is used so often that one is prone to forget its significance. One first immediate consequence is the formula for the sum of the (interior) angles of a convex polygon. If it has n vertices, it can be dissected into $n - 2$ triangles and has $(n - 2)2R$ (or, expressed in another way, $(n - 2)\pi$) as the sum of its angles. Proclus proves this in his comment on I.32. (Proclus–Morrow p. 301.) Proclus proceeds (p. 302) to state that "the property of having its interior angles equal to two right angles is an essential property of the triangle as such." He refers to Aristotle for the meaning of "essential property." In the words of today this means that the triangle is characterized (among convex polygons) by the sum of its angles: A convex polygon is a triangle if and only if the sum of its interior angles equals two right angles. Theorem I.32 has played its role in philosophy later on as well. For Immanuel Kant it is the quintessential example of what he calls "a synthetic a priori judgment," that is, a statement of absolute certainty (not depending on experience) that adds to our knowledge (*Critique of Pure Reason* B 744–746).

One of the far-reaching consequences of I.32 was found by Jacob Steiner (1796–1863). He used the formula $(n - 2)\pi$ for the sum of the interior angles of a polygon for a simple proof of Euler's formula for convex polyhedra: If such a polyhedron has v vertices, e edges, and f faces, then "invariably"

$$v - e + f = 2.$$

Thus the simple invariant of triangles goes as far as proving one of the most important invariants of modern algebraic topology, the Euler characteristic, in its first significant case of convex polyhedra.

4.5   Book I, Part C: Parallelograms and Their Areas

In part C we find a systematic study of the interrelations between the concepts of "parallelism" and "of equal content."

Euclid defines various types of "quadrilaterial figures" in Def. 22 at the beginning of Book I, but not the parallelograms that figure so prominently in this section C. Instead, he introduces them together with their basic symmetry properties in Propositions 33 and 34.

Prop. I.33.
The straight lines joining equal and parallel straight lines (at the extremities which are) in the same directions (respectively) are themselves equal and parallel.

Prop. I.34.
In parallelogrammic areas the opposite sides and angles are equal to one another, and the diameter bisects the areas.

In Prop. 34 Euclid speaks about halving the "area" of a parallelogram, but he does not use this word in the subsequent propositions, which are—in our understanding—equally statements about areas. In daily life, the Greeks measured their properties, and in fact the very word "geometry" means "measuring the fields." Measuring a field means attaching a number to it; it measures so and so many square feet. In mathematical language this amounts to a function that associates numbers to certain (polygonic) plane surfaces. But the concept of a function is alien to the *Elements*. Euclid does not use it, and moreover, he does not use any formulas that in effect would define functions. For a modern description of what Euclid does, we quote Hartshorne, Ch. I.3, "The theory of area," about Euclid's notion of "equal figures":

> So what did Euclid have in mind? Since he does not define it, we will consider this new equality as an undefined notion, just as the notions of congruence for line segments and angles were undefined. We will call this new notion equal content, to avoid confusion with other notions of equality or congruence. We do not want to use the word area, because this notion is quite different from our common understanding of area as a function associating a number to each figure.
>
> From the way Euclid treats this notion, it is clear that he regards it as an equivalence relation, satisfying the common notions. In particular
>
> (a) Congruent figures have equal content.

(b) If two figures each have equal content with a third, they have equal content.

(c) If pairs of figures with equal content are added in the sense of being joined without overlap to make bigger figures, then these added figures have equal content.

(d) Ditto for subtraction, noting that equality of content of the difference does not depend on where the equal pieces were removed.

(e) Halves of figures of equal content have equal content (used in the proof of I.37).

(f) The whole is greater than the part, which in this case means if one figure is properly contained in another, then the two figures cannot have equal content (used in the proof of I.39).

In terms of the axiomatic development of the subject, at this point Euclid is introducing a new undefined relation, and taking all the properties just listed as new axioms governing this new relation.

In the next propositions, 35–41, Euclid achieves more flexibility in handling the concept of equal content, or equality, as he says.

Prop. I.35.
Parallelograms which are on the same base and in the same parallels are equal to one another.

Prop. I.36.
Parallelograms which are on equal bases and in the same parallels are equal to one another [Fig. 4.15].

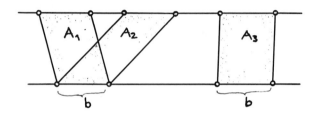

FIGURE 4.15

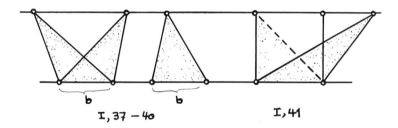

FIGURE 4.16

Propositions 37–40 say similar things for triangles, and Proposition 41 associates parallelograms and triangles. (The parallelogram situated as in Fig. 4.16 is the double of the triangle.)

At this point the theory of equal content branches out in two directions. The first branch leads directly to the theorem of Pythagoras (I.46–48), which in any case is a goal in its own right; and the second one leads via I.42–45 and the theorem of Pythagoras to the important result II.14: *It is possible to construct a square of content equal to that of any rectilinear figure.* Or shorter: Any rectilinear figure can be squared.

We stick to Euclid's sequence and discuss I.42–45, which will find their sequels in Book II.

Prop. I.42.
To construct, in a given rectilineal angle, a parallelogram equal to a given triangle.

The construction is easy enough, compare Fig. 4.17, where $\triangle ABC$ and the angle δ are given and E is the midpoint of BC.

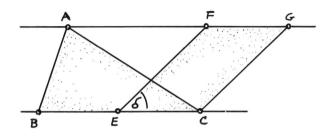

FIGURE 4.17

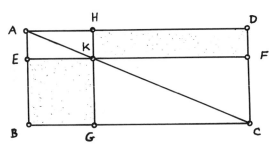

FIGURE 4.18

The "given angle" will be a right angle in Euclid's subsequent applications. So we might as well specialize it to this case in the next propositions. (The generalization from rectangles to parallelograms is easy enough because shears preserve areas.) We replace "parallelogram" by "rectangle" and "given angle" by "right angle" in Euclid's Props. I.43–45. Book II is about rectangles throughout.

Prop. I.43.

In any rectangle the complements of the rectangles about the diagonal are equal to one another.

Figure 4.18 shows a diagram that is used over and over again in the *Elements*. In several contexts Euclid simply calls it "the schema." The point K is on the diagonal of the rectangle □$ABCD$, and the lines EF, GH are parallel to the sides. Euclid denotes □$BGKE$ by BK, and □$KFDH$ by KD. These latter rectangles are the "so-called complements." (The reader may want to look ahead at Props. VI.16, 24, 26.)

We have to prove:

K is on $AC \Longrightarrow$ □BK and □KD are of equal content (are equal).

By I.34, the triangle $\triangle ABC$ is equal to $\triangle DAC$. For the same reason, $\triangle GCK$ and $\triangle FKC$ as well as $\triangle AEK$ and $\triangle KHA$ are equal. Subtracting the two smaller triangles from the large one on each side of the diagonal gives the result.

Simple as it is, Prop. 43 has very many useful consequences. The next proposition is the first one. By "applying" a figure C to a line (segment) AB Euclid means to construct a rectangle with one side AB of equal content with figure C.

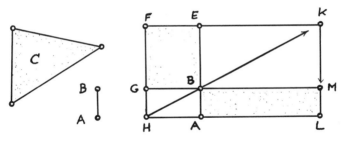

FIGURE 4.19

Prop. I.44.
To a given straight line to apply a rectangle equal to a given triangle.

Construction. Let △C and line AB be the given figures (Fig. 4.19). Construct a rectangle ▭BF of equal content with △C via I.42. Place AB so that it prolongs side EB and construct ▭BH. Prolong FE and HB until they meet in K. (Euclid shows that they will meet by means of Post. 5.) Complete the figure as shown in Fig. 4.19. ▭BL has one side AB and is of equal content with ▭BF by Prop. I.43.

Prop. I.45.
To construct a rectangle equal to a given rectilinear figure.

For the given figure Euclid takes a quadrangle, dissects it into two triangles, and transforms these by Prop. I.44 into two rectangles with a common side. By combining the two rectangles with the common side he gets one rectangle as desired. The proof is done meticulously by justifying every single step.

In spite of the general assertion Euclid, selects a quadrangle for the proof. But the procedure is quite transparent, and it is obvious how to proceed in the general case. This way of handling proofs, which today might be done by mathematical induction, is quite typical for Euclid. We will see it now and then on other occasions.

4.5.1 Comment on Props. I.44/45

We will for a moment use modern formulas. The area A of a rectangle with sides (of length) a, b is given by $A = ab$. In I.44, let R be the given rectangle and a be the given side. In terms of these formulas, the problem of I.45 amounts to the solution of the linear equation

$$R = ax,$$

where x is the second side of the desired new rectangle. Seen this way, I.45 is algebra in geometric disguise, and hence it has been interpreted as "geometric algebra." Historians have said that this interpretation is not justified and an anachronism. Mathematicians have replied that the formulas represent an isomorphic image of the geometric situation and hence are the correct modern way of describing Euclid's procedures. The same problem arises again in Book VI, where the geometric equivalent of quadratic equations is treated. Because the geometric version is quite sufficient for the understanding of Euclid's text, we will leave the formulas aside. Occasionally we will use them in order to facilitate understanding for the modern reader.

4.6   Book I, Part D: The Theorem of Pythagoras

In proposition I.46 Euclid shows how to construct a square on a given line; I.47 is the famous theorem of Pythagoras, and I.48 is its converse. We quote I.47 and its proof verbatim.

Prop. I.47.
In right-angled triangles the square on the side subtending the right angle is equal to the squares on the sides containing the right angle.

Let ABC be a right angled triangle having the angle BAC right;
I say that the square on BC is equal to the squares on BA, AC.
For let there be described on BC the square $BDEC$, and on BA, AC the squares GB, HC; through A let AL be drawn parallel to either BD or CE, and let AD, FC be joined [Fig. 4.20].
Then, since each of the angles BAC, BAG is right, it follows that with a straight line BA, and at the point A on it, the two straight lines AC, AG not lying on the same side make the adjacent angles equal to two right angles;

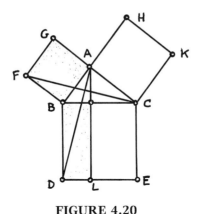

FIGURE 4.20

therefore CA is in a straight line with AG.
For the same reason

BA is also in a straight line with AH.

And, since the angle DBC is equal to the angle FBA: for each is right: let the angle ABC be added to each;
therefore the whole angle DBA is equal to the whole angle FBC.
And, since DB is equal to BC, and FB to BA, the two sides AB, BD are equal to the two sides FB, BC respectively, and the angle ABD is equal to the angle FBC;
therefore the base AD is equal to the base FC, and the triangle ABD is equal to the triangle FBC.   [I.4]
Now the parallelogram BL is double of the triangle ABD, for they have the same base BD and are in the same parallel BD, AL.   [I.4I]
And the square GB is double of the triangle FBC, for they again have the same base FB and are in the same parallels FB, GC.   [I. 4I]
Therefore the parallelogram BL is also equal to the square GB.
Similarly, if AE, BK are joined, the parallelogram CL can also be proved equal to the square HC;
therefore the whole square $BDEC$ is equal to the two squares GB, HC.
And the square $BDEC$ is described on BC, and the squares GB, HC on BA, AC.
Therefore the square on the side BC is equal to the squares on the sides BA, AC.

Let us recapitulate the main points of the proof. The right angle at A guarantees that G, A, C are in a straight line parallel to FB. This is the decisive point. For the equality of $\square GB$ and $\square BL$ Euclid has to resort to their respective halves, $\triangle FBA$ (which is not shown) and $\triangle BDL$ (also not shown). By I.41 these are equal to (i.e., are of equal content with) $\triangle FBC$ and $\triangle BDA$. These two triangles are congruent by SAS, and we are done.

Comment There are many dozens of different proofs of Pythagoras's theorem. Proclus credits Euclid personally with this one. It is a marvellous piece of mathematics, and I personally like it better than any other proof. There is no special trick or need of a formula, one sees in such a clear way how the square $\square GB$ is transformed into the rectangle $\square BL$, and in spite of its simplicity the argument is in no way trivial.

The theorem of Pythagoras is as fundamental for mathematics today as it was in Euclid's time. It is the progenitor of all the different kinds of metrics and of quadratic forms, and of theorems like $\sin^2\alpha + \cos^2\alpha = 1$. Via its generalization, the law of cosines, and the corresponding scalar product in vector spaces, it pervades mathematics as far as the eye can see.

Prop. I.48

If in a triangle the square on one of the sides is equal to the squares on the remaining two sides of the triangle, the angle contained by the remaining two sides of the triangle is right.

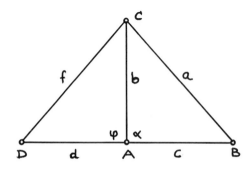

FIGURE 4.21

Proof

Let $\triangle ABC$ be the given triangle. We take the notation from Fig. 4.21 and abbreviate. Let AD be equal to AB and φ be a right angle, then by I.47 we have $f^2 = d^2 + b^2 = c^2 + b^2$, which by assumption is equal to a^2. Hence f is equal to a. (Here is a little gap. This implication has not been proved before.) Now, by the congruence theorem SSS the two triangles $\triangle ABC$ and $\triangle ADC$ are congruent, and hence $\alpha = \varphi$ is a right angle.

Propositions I.47 and 48 combined are the full theorem of Pythagoras.

We conclude this chapter by quoting a fine sonnet by the German poet Adelbert von Chamisso, translated by Max Delbrück, together with a nice remark by C. L. Dodgson. According to a legend from antiquity, Pythagoras sacrificed a hundred (a hecatomb, or, in another version, only one) oxen to the gods after he had discovered his theorem. (The German original of the poem can be found in the notes.)

Adelbert von Chamisso: The Truth

(Translated by Max Delbrück)

The TRUTH: her hallmark is ETERNITY
As soon as stupid world has seen her light
PYTHAGORAS' theorem today is just as right
As when it first was shown to the FRATERNITY.

The GODS who sent to him this ray of light
to them PYTHAGORAS a token sacrificed:
One hundred oxen, roasted, cut, and sliced
Expressed his thank to them, to their delight.

The oxen, since that day, when they surmise
That a new truth may be unveiling
Forthwith burst forth in fiendish railing.

PYTHAGORAS forever gives them jitters –
Quite powerless to stem the thrust of such emitters
of LIGHT, they tremble and they close their eyes.

But neither thirty years, not thirty centuries, affect the clearness, or the charm, of Geometrical truths. Such a theorem as "the square of the hypotenuse of a rightangled triangle is equal to the sum of the squares of the sides" is as dazzlingly beautiful now as it was in the day when Pythagoras first discovered it, and celebrated its advent, it is said, by sacrificing a hecatomb of oxen – a method of doing honor to Science that has always seemed to me *slightly* exaggerated and uncalled-for. One can imagine oneself, even in these degenerate days, marking the epoch of some brilliant scientific discovery by inviting a convivial friend or two, to join one in a beefsteak and a bottle of wine. But a *hecatomb* of oxen! It would produce a quite inconvenient supply of beef.

C. L. Dodgson (Lewis Carroll)

5

The Origin of Mathematics 2

Parallels and Axioms

....

The discussion of the parallel axiom has been the driving force behind the axiomatization of mathematics. We will sketch the development in the history of mathematics. (For more detailed information see the Notes.)

Historians generally assume that the introduction of Post. 5 and the theory of parallels was fairly recent at Euclid's time. Before that, there may have been a definition of parallels comprising properties like the definition I.17 of the diameter of a circle: "A diameter of the circle is a straight line drawn through the center ... and such a straight line also bisects the circle." A similar definition of parallels may have said that they exist, are unique, and can be constructed using alternate angles, or more simply, right angles. (Possibly even their property of being equidistant was stated.) These aspects of parallels were quite natural for architects and stone masons, who worked with parallel layers of stones all the time and were used to very precise measurements.

A faint hint of the origin of the parallel axiom (Post. 5 for short in the following text) may be a remark by Proclus in his discussion of

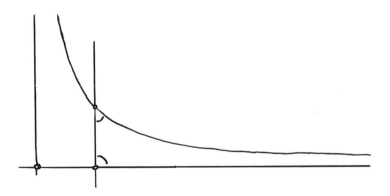

FIGURE 5.1

Post. 5: "That there are lines that approach each other indefinitely but never meet seems implausible and paradoxical, yet it is nevertheless true and has been ascertained for other species of lines" (Proclus–Morrow 151). The most prominent example of such lines are a hyperbola and its asymptotes (Fig. 5.1). Admitting angles between curved (and straight) lines, this is a "counterexample" to Post. 5, or at least it could draw the attention of mathematicians to the problem.

The first to write about conics was Menaechmus, who may have lived about 380–320 B.C.E. If the speculation above is right, this would put the introduction of Post. 5 at a time around 340. Other historians think it may have been introduced by Euclid himself. Aristotle never mentions the postulates of Book I.

Compared to the other postulates, the parallel postulate is a rather complicated statement and not as obvious as, say, the equality of all right angles. Therefore, mathematicians of antiquity tried to eliminate it, either by proving it or by replacing it by a simpler and more intuitively convincing axiom. However, none of them succeeded.

The problem was taken up again in the seventeenth century. John Wallis gave a public talk in Oxford in 1663 about the subject and proved the following: If there are similar triangles of different content, then Post. 5 is true. This result replaces Post. 5 by a more plausible one. Several other mathematicians tried the following way: Start with the negation of Post. 5 and try to find consequences that contradict established theorems. The most prominent among these was the Jesuit Girolamo Saccheri, who in 1733 published a book entitled *Euclid liberated from every defect*. Although it was largely forgotten, this book contained a key distinction. Saccheri looked at a quadrangle that was constructed with 3 right angles. Then there are three alternatives for the remaining fourth angle: It might be obtuse, right, or acute. The hypothesis of the right angle is equivalent to Post. 5. Then he managed to find a contradiction derived from the hypothesis of the obtuse angle. (It contradicted the existence of lines of infinite length. This is the situation on the sphere.)

From the hypothesis of the acute angle he drew many consequences but could not find a contradiction. (He believed he had one, but there was a mistake.) In 1766 Johann Heinrich Lambert wrote a paper on "The Theory of Parallel Lines," which was published posthumously in 1786. It seems that he thought about a valid geometry that could be derived from Saccheri's hypothesis of the acute angle. He points to Euclid's I.16/17 as evidence that Euclid held similar opinions.

Finally, around 1830 three mathematicians were thoroughly convinced of the existence of a valid "non-Euclidean" geometry: Carl Friedrich Gauss, János Bolyai, and Nikolai Ivanovitch Lobachevsky. But Gauss did not publish what he had found, and the papers by Bolyai (1832) and Lobachevsky (1835) were hard to read. So the subject lay dormant until the second half of the 1860s, when Gauss's (who had died in 1855) private letters were published and the situation changed dramatically. Some papers by Riemann on abstract geometries in the 1850s, the development of differential geometry, and the prestige of Gauss made non-Euclidean geometry one of the hottest research topics in mathematics. The theory derived from the "hypothesis of the acute angle" was now called "hyperbolic geometry." The possibility of a (still) hidden contradiction was ruled out by Beltrami (1868), who interpreted it as the geometry on a surface with constant negative curvature. Felix Klein (1871) found the now so-called Cayley–Klein model in projective geometry, and in 1882 Poincaré placed it in the context of complex function theory. The question of the parallel axiom was finally settled by 1880: There are three types of plane geometry that satisfy Euclid's other axioms: elliptic geometry (on the sphere, with antipodal points identified, the hypothesis of the obtuse angle) with no parallels at all; Euclidean geometry with Post. 5 (hypothesis of the right angle), where parallels exist and are unique; and hyperbolic geometry (hypothesis of the acute angle), where parallels exist but are not unique. In elliptic geometry the sum of the angles in a triangle is greater then 180°, in Euclidean geometry it is 180°, and in hyperbolic geometry it is smaller than 180°.

But there is more to the story of axioms in geometry. Euclid had employed the notions of ordering and betweenness of points on a line intuitively. In effect, he had ruled out the hypothesis of the obtuse angle by these means. The subject of ordering was taken up by Moritz Pasch in a book *Lectures on Recent Geometry* (*Vorlesungen über neuere Geometrie*) in 1882. His axioms about betweenness essentially completed the axiomatization of plane geometry. (Pasch had such an acute sense for fine points in logic that in his later days he was the foremost expert on the bylaws of his university in Giessen. His son-in-law C. Thaer produced the standard German translation of Euclid.)

For Pasch, geometry was still the science of physical space. This last barrier was broken by David Hilbert (1899) in his *Foundations of Geometry* (*Grundlagen der Geometrie*). Hilbert explicitly says that the objects of geometry are called points, lines, planes by convention, but they could be called by other, fancy, names just as well. These objects are defined only "implicitly" by what is said about them in the axioms. (For instance, a point might be a pair of real numbers.) The subject of geometry has changed from a study of space to the study of the logical interdependence of certain statements about otherwise undefined objects.

Hilbert grouped the axioms under five headings: axioms of incidence like "Two different points lie on a unique line," axioms of ordering, axioms of congruence, the parallel axiom, and finally, axioms of continuity.

The axioms of continuity guarantee that the points of a line may be identified with the real numbers. This enables him to prove his main theorem: *The five groups of axioms determine the Euclidean plane (up to isomorphism) uniquely. It may be viewed as the plane of analytic geometry over the field of real numbers.*

In an appendix to his book Hilbert presents the first axiom system for the real numbers. Starting from this point, the axiomatic method conquered the whole of mathematics in the twentieth century.

The Origin of Mathematics 3

Pythagoras of Samos

Pythagoras lived about 570–490 B.C.E. The only roughly determined date in his life is ≈ 530, when he left Samos to settle in Crotona, in southern Italy. At Crotona he founded a religious and philosophical society that soon came to exert considerable political influence in the Greek cities of southern Italy. He was forced to leave Crotona about 500 and retired to Metapontum, where he died (see Fig. 6.1).

The Pythagoreans, as his followers were called, continued to exert political power until sometime in the middle or late fifth century, when a democratic revolution occurred and they were forced to leave the Greek cities of southern Italy. Some of them went to Sicily and others to the Greek mainland, where they found new centers for their activities. The last of the Pythagoreans were known in about 350 B.C.E. as poor vegetarian wandering pilgrims.

The city (and island) of Samos together with its close neighbors Miletus and Ephesus on what is now the Turkish coast were booming economic and intellectual centers in the sixth century B.C.E. Thales and his student Anaximandros taught in Miletus in the first half

51

FIGURE 6.1 (a) Coin from Metapontum with a "pentagram" composed of grains of barley (about 440 B.C.E.?). (b) Coin from Abdera (430 B.C.E.?) showing an idealized portrait of (Π) *YΘAΓOPHΣ*, that is, (P)YTHAGORES. (c) Coin from Melos (before 420 B.C.E.) with pentagram

of the sixth century. The philosopher Heraclitus of Ephesus was a contemporary of Pythagoras.

Two examples of practical geometry will illustrate the atmosphere of the city of Pythagoras's youth. Outside of Samos was the ancient sanctuary of the goddess Hera. In 570–560 the city commissioned the architects Rhoikos and Theodoros with the construction of a new temple of dimensions hitherto unheard of. Its ground plan was 100 × 200 cubits (52.5 × 105 meters); its 104 columns were 18 meters high. The bases of the columns had a diameter of up to 1.80 meters and weighed about 1500 kg each. In spite of these extraordinary dimensions, they were turned on lathes! This temple was to be the prototype of the Ionian style in architecture. It was constructed during Pythagoras's youth. Just completed, it was destroyed in a revolution in the 530s, which brought the tyrant Polykrates into power. Polykrates immediately ordered the construction of a new and even bigger temple.

The second example of high technology in the second half of the sixth century in Samos is the tunnel of Eupalinus. This tunnel, for a new water main, was about 1 km long and, this is the noteworthy point, was built from both sides of the mountain! The diggers met in the middle of the mountain with a deviation of about 10 meters. The people who gave the money for the construction did indeed trust geometry and geodesics.

Such was the background of Pythagoras's youth. He may have traveled to Egypt and Babylon for studies, but Samos was where the action was in his time.

a   b

FIGURE 6.2 (a) Tetraktys after Iamblichus. (α stands for the unit one.) (b) Pentagram from St. Mary's Church, in Lemgo, Germany (1300 C.E.).

The Mathematics of the Pythagoreans

Almost nothing is known for sure about Pythagoras's own mathematical achievements. The oral tradition of his followers was written down rather late, some of it by the students of Aristotle. The best available information about Pythagoras is collected in the book by Burkert [1972]. It seems that the students of Pythagoras in southern Italy were split into two groups: the Akusmatikoi, who followed the master's sayings verbatim, and the Mathematikoi, who tried to develop and expand his teachings. The Pythagoreans used two symbols: the so-called tetraktys, the arrangement of ten dots representing $1 + 2 + 3 + 4 = 10$, and the pentagram (Fig. 6.2).

They used to call the pentagram "Health." (There is indeed a very old tradition of the pentagram as a medical symbol, cf. Notes.) Undoubtedly, musical harmonics was central to Pythagoras's teachings. The Middle Ages knew him as "inventor musicae," not as a mathematician. Aristotle writes about the Pythagoreans:

> Contemporaneously with these philosophers and before them, the Pythagoreans, as they are called, devoted themselves to mathematics; they were the first to advance this study, and having been brought up in it they thought its principles were the principles of all things. Since of these principles numbers are by nature the first, and in numbers they seemed to see many resem-

blances to the things that exist and come into being—more than
in fire and earth and water (such and such a modification of numbers being justice, another being soul and reason, another being
opportunity—and similarly almost all other things being numer-
ically expressible); since, again, they saw that the attributes and
the ratios of the musical scales were expressible in numbers; since,
then, all other things seemed in their whole nature to be modelled
after numbers, and numbers seemed to be the first things in the
whole of nature, they supposed the elements of numbers to be
the elements of all things, and the whole heaven to be a musical
scale and a number. And all the properties of numbers and scales
which they could show to agree with the attributes and parts and
the whole arrangement of the heavens, they collected and fitted
into their scheme; and if there was a gap anywhere, they readily made additions so as to make their whole theory coherent.
(*Metaphysics* 985 b 23–986 a 7)

In the first part of this quotation, Aristotle praises the Pythagoreans for their mathematical studies. In the second part, he criticizes
them for unfounded speculations. It seems that there were four
"mathematical" parts of the Pythagorean teachings: arithmetic, geometry, harmonics (music), and astronomy. This is the classical
"quadrivium," part of the seven liberal arts. In the Middle Ages, the
three "trivial" parts, the "trivium," grammar, rhetoric, dialectics, were
the introductory ones of the seven undergraduate courses. Plato was
the first to call the subjects of the quadrivium "mathemata," things
to be learned. (*Republic* 527 c 10)

Arithmetic

Aristotle points out the prominent role of numbers in Pythagorean
mathematics. There are considerable traces of Pythagorean arithmetic in the writings of Nicomachus, which we will present and
contrast to Euclid's style in an appendix to the arithmetical Book VII
of Euclid.

The Application of Areas

We have seen the first example of the application of areas in I.44, the simple application of an area to a line. The next, and more substantial, examples would be II.5,6 when supplemented by the theorem of Pythagoras so as to give the solution of a quadratic problem. (Readers should compare the more detailed statements and comments in the discussion of Book II.) It reads like this. Let an area c^2 and a line b be given. Find a line x such that

$$bx - x^2 = c^2 \qquad \text{(application with square defect)}$$

or

$$bx + x^2 = c^2 \qquad \text{(application with square excess)}.$$

The geometric equivalent proceeds like this. (We take the case with quadratic excess.) By II.14, the given area might be a square. By II.6, we are able to transform the problem into the geometric version of

$$\left(x + \frac{b}{2}\right)^2 = x^2 + bx + \left(\frac{b}{2}\right)^2 = c^2 + \left(\frac{b}{2}\right)^2.$$

Pythagoras's theorem then provides us with z such that

$$c^2 + \left(\frac{b}{2}\right)^2 = z^2,$$

and from this we find x as in Fig. 6.3. Proclus ascribes the discovery of this technique to the Pythagoreans in his comment on *Elements* I.46:

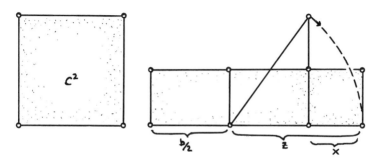

FIGURE 6.3

Eudemus and his school tell us that these things—the application of areas, their exceeding and their falling short—are ancient discoveries of the Pythagorean muse.... Those godlike men of old saw the significance of these terms in the describing of plane areas along a finite straight line. (Proclus–Morrow p. 332)

Incommensurable Segments

Incommensurable segments, or, as we might say today, segments of irrational length with respect to a given unit segment, are one of the most important discoveries of the Greek mathematicians. Pappus tells us that the theory of incommensurable lines "had its origin in the school of Pythagoras" (Heath [1921], 154). The history and significance of this subject will be discussed in an appendix to Book X.

The Dodecahedron

One of the "mathematikoi" students of Pythagoras was Hippasus of Metapontum. Sources from late antiquity report about him that he

was a Pythagorean but, owing to his being the first to publish and write down the (construction of the) sphere with the twelve pentagons, perished by shipwreck for his impiety, but received credit for the discovery. (Heath [1921], 160 after Iamblichus)

Again there will be a special appendix dealing with the regular polyhedra and their history following Book XIII.

The Pythagorean Theorem as a Paradigm for Mathematics

The most frequent answer to the question of what people who are disconnected from mathematics in their adult life remember of school mathematics is "Pythagoras." And I think that this is as it should be. I would rank the Pythagorean theorem as a cultural asset of the first order, the knowledge of which should be bestowed upon every student as a standard for life. It belongs to a basis of common intellectual possessions of mankind. In Greek antiquity, cultural coherence rested on the general widespread intensive knowledge of Homer. In the Middle Ages the Bible (besides the Latin language) had a similar function in Western Europe. More recently, in the English-speaking world one recognizes cultural coherence in that an allusion to Shakespeares's *Hamlet* is immediately understood. (In German, Goethe's *Faust* plays an analogous role.) Mathematics, however, is independent of specific languages and transcends cultural borders. The theorem of Pythagoras is taught and understood all over the world. And it is more important than pop-music.

An approach to the theorem of Pythagoras through a work of art will help us to understand its significance for mathematics as a whole.

When I saw from afar a sculpture by the german sculptor Helmut Lander at an exhibition my spontaneous reaction was, "Pythagoras at work on the square on the side." (See Fig. 6.4.)

The right-angled triangle and the square block lead a mathematician immediately to the association "Pythagoras," but the artist himself chose "Sisyphus" as the title of his work. Can we adhere to "Pythagoras," beyond the external similarity, and argue with the sculptor about his title?

A few remarks to this purpose about the Pythagorean theorem. One indeed knows the figure, but the assertion of the theorem does remain invisible. The equality of the areas expressed in $a^2 + b^2 = c^2$ cannot be seen from the figure. Entirely different from the case of the intersection of the three altitudes of a triangle, knowledge here eludes the direct view. Only the proof gives a reason for believing

FIGURE 6.4

and insight into the truth of the theorem. And that is exactly what one calls "deep" in mathematics.

Here something is not just verified, or something that in any case could be read off from a good drawing; it is placed within the framework of logical deduction. The theorem of Pythagoras is one of the few important examples of substantial mathematics that is usually treated in school. Quite rightly does it often stand as a symbol for all of mathematics. It is a good sign for the intellectual perception of many nonmathematicians if it is exactly this theorem that they remember from their schooldays. Part of the symbol is also the laborious way to knowledge and the drudgery with formulas and proofs that on the boundaries of research is the same as initiation in high school.

Though to be sure, with some real distinctions—which brings us back to the sculpture: In research there is no longer a teacher who knows the way and has the solution ready. One works hard and often, again and again, in vain. One believes that one has just mastered the problem when the whole structure breaks down again at some decisive point. And if one finally manages to put everything together successfully, every answer simply generates new questions. The game starts all over again.

Thereby we would be with Sisyphus, who in never-ending effort pushes his block up the mountain. Our modern image of this ancient figure is really stamped with the existential meaning that Albert Camus has given him: the heroic man who consciously takes upon himself the contradiction of life. To him corresponds the formal figuration of the sculpture. Exerting all its power, the organic form asserts itself between the overpowering abstract figures of the triangle and the square. It is lost if the blocks slam together: In the geometrically precise conception there is no place for human beings.

And so it is with the Pythagorean theorem: When the assertion is ready, it stands there in cool precision. The human element enters in the proof where the squares on the sides must be transformed and reformed until they come together again on the hypotenuse.

So now should one say "Pythagoras" or "Sisyphus"? Lander himself says it doesn't matter. The figure forms a bridge between the proverbial two cultures of science and the humanities (where mathematics, in the sense of this classification, belongs to the sciences, even if she should protest against it). I plead for Pythagoras. Naming it so would be more unconventional and would provoke more association and further thought. The starting point becomes more specific, and the viewer is spoken to more personally than with the rather philosophical Sisyphus, who then appears already alone.

7

C H A P T E R

Euclid Book II

The Geometry of Rectangles

Book II is short and homogeneous, with only 14 propositions and two definitions at the beginning. For the most part it is about various combinations of rectangles and squares of equal content. At the end we find the generalization of the theorem of Pythagoras to what today is called the law of cosines and, as the last proposition, the squaring of a rectangle. There are reasons to assume a Pythagorean origin for the main part of Book II, but some historians have different opinions.

We will follow Euclid and speak of "equal" rectangles where we could use "of equal content" in the modern way.

II. Def. 1. *Any rectangle is said to be contained by the straight lines containing the right angle.*

II. Def. 2. *(For rectangles) Let any one of the rectangles around the diameter of any rectangular area (together) with the two complements be called a gnomon.*

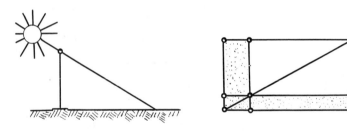

FIGURE 7.1   Gnomons

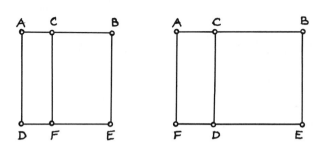

FIGURE 7.2

A gnomon was originally a sort of primitive sundial, a pole perpendicular to the horizon whose shadow was used for the measurement of time. The mathematicians apparently transferred the name to the similar-looking geometric figure. Figure 7.1 shows both of them.

We omit Prop. II.1, which is a generalization of II.2/3 and has probably been interpolated after Euclid. For II.2/3, a figure (Fig. 7.2) and a statement in shorthand suffice.

II.2: $\square AE = \square DC + \square FB$
II.3: $\square AE = \square AD + \square BD$

In the first case, we might say that we see a square with defect, namely $\square DC$, and in the second one we have a square with excess $\square AD$. The technical terms "defect" and "excess" will reappear later.

Prop. II.4.
If a straight line is cut at random, the square on the whole is equal to the squares on the segments and twice the rectangle contained by the segments.

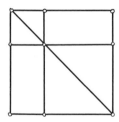 

FIGURE 7.3

For the modern reader, this, together with the figure, is the familiar geometric interpretation of the binomial formula for $(a+b)^2$. For Euclid, however, it is a purely geometrical statement. (He does, however, apply the equally geometric Prop. II.6 to numbers in the lemma to Prop. X.28.) The proof is not difficult, but straightforward and somewhat lengthy.

Curiously enough, Prop. II.4 has found its way to a Greek coin (Fig. 7.3). From the year 404 B.C.E. onwards, the reverse of the coins of the city (and island) of Aigina, not far from Athens, shows the design of Euclid's Prop. II.4. We do not know why the people of Aigina selected this particular design for their money. But in any case the coins indicate the familiarity of Prop. II.4 some hundred years before Euclid's time.

Prop. II.5.

If a straight line be cut into equal and unequal segments, the rectangle contained by the unequal segments of the whole together with the square on the straight line between the points of section is equal to the square on the half.

For let a straight line AB be cut into equal segments at C and into unequal segments at D [Fig. 7.4];

I say that the rectangle contained by AD, DB together with the square on CD is equal to the square on CB.

The proposition is true because $\square AL$ and $\square DF$ are of equal content.

The next proposition is very similar. We quote it in the shorter version before we comment on both of them.

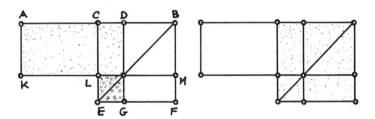

FIGURE 7.4

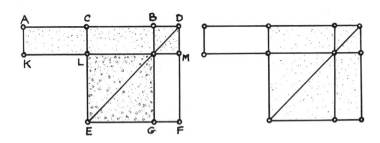

FIGURE 7.5

Prop. II.6.

Let a straight line AB be bisected at the point C, and let a straight line BD be added to it in a straight line. I say that the rectangle contained by AD, DB together with the square on CB is equal to the square on CD. (See Fig. 7.5.)

Again the truth of the proposition is easily seen from the figure.

Comments on II.5/6.

The last two propositions have often been associated with quadratic equations in the following way. In II.5, let $AB = b$ and $DB = x$. Then II.5 translates into the algebraic formula

$$\left(\frac{b}{2}\right)^2 = \left(\frac{b}{2} - x\right)^2 + x(b - x),$$

or equivalently,

$$x^2 - bx + \left(\frac{b}{2}\right)^2 = \left(\frac{b}{2} - x\right)^2.$$

Similarly, for II.6 with $AB = b$ and $BD = x$,

$$(b + x)x + \left(\frac{b}{2}\right)^2 = \left(\frac{b}{2} + x\right)^2.$$

That is, II.5/6 show us, in the literal sense, how to complete the square. Euclid had to consider these two cases separately because there are no negative segments or areas. Completing the square is the first step in solving a quadratic equation like

$$x^2 + bx + c = 0,$$

$$x^2 + bx + \left(\frac{b}{2}\right)^2 = \left(\frac{b}{2}\right)^2 - c,$$

$$\left(x + \frac{b}{2}\right)^2 = \left(\frac{b}{2}\right)^2 - c.$$

For the next step, one has to extract the square root out of $\left(\frac{b}{2}\right)^2 - c$. This depends on the size and sign of c and hence would again need special consideration for Euclid. He executes this second step in Book VI, somewhat hidden in VI.25, but not in Book II, even if it would be easy enough with the help of Pythagoras's theorem. Therefore one cannot speak of II.5/6 as "solutions to quadratic problems." They can be used only in order to verify known solutions and not in order to find unknown solutions. This interpretation is confirmed by the fact that Euclid introduces the condition $\left(\frac{b}{2}\right)^2 > c$, which is necessary for the existence of a solution, only in Props. VI.27/28.

Let us add another translation of II.5. Let $AC = a$ and $CD = b$. Then $AD = a + b$ and $BD = a - b$, and hence by II.5,

$$(a + b)(a - b) + b^2 = a^2, \tag{7.1}$$

or

$$(a + b)(a - b) = a^2 - b^2. \tag{7.2}$$

The next propositions, II.7–10, are variations on the themes of II.4–6. After these, Euclid turns to three different subjects. II.11 is basic for the construction of the regular pentagon in Book IV; Propositions II.12, 13 generalize the theorem of Pythagoras; and II.14 is the capstone of the sequence, dealing with the squaring of polygons.

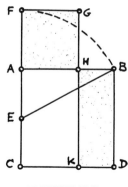

FIGURE 7.6

Prop. II.11.

To cut a given straight line so that the rectangle contained by the whole and one of the segments is equal to the square on the remaining segment.

Let the given line be AB. We are looking for a point H on AB such that the rectangle contained by AB and HB is equal to the square on AH.

Euclid first describes the construction of the accompanying figure (Fig. 7.6):

Let the square $ABDC$ be described on AB;
let AC be bisected at the point E, and let BE be joined;
let CA be drawn through to F, and let EF be made equal to BE;
let the square FH be described on AF, and let GH be drawn through to K.

With the help of II.6 and the theorem of Pythagoras I.47 he shows that the point H has the desired properties:

On the line CF we have the situation of II.6: AC has been bisected in E, and AF has been added to it; hence

$$\square(CF, FA) + \square AE = \square EF.$$

By construction, $\square EF = \square EB$, and by I.47, $\square EB = \square AE + \square AB$, hence

$$\square(CF, FA) + \square AE = \square AE + \square AB.$$

This implies

$$\square(CF, FA) = \square AB.$$

Subtraction of $\square(AC, AH)$ on both sides gives us the desired result:

$$\square AH = \square AF = \square(BD, HB) = \square(AB, BH).$$

Comment.

Typically, Euclid does not tell us how he arrived at his construction, but only verifies his solution. This is in accordance with the remarks about II.5/6. Later on, in VI.30, he unveils the secret. Let us again use modern notation as a convenient shorthand. Let b be the given segment and x the larger part of it. We have the condition

$$b(b - x) = x^2, \tag{7.3}$$
$$b^2 = x^2 + bx. \tag{7.4}$$

With II.6 this is transformed into

$$b^2 + \left(\frac{b}{2}\right)^2 = \left(x + \frac{b}{2}\right)^2,$$

and the theorem of Pythagoras provides us with the segment $EB = d$ such that

$$d^2 = b^2 + \left(\frac{b}{2}\right)^2 = \left(x + \frac{b}{2}\right)^2;$$

hence

$$x = d - \frac{b}{2}, \text{ as constructed.}$$

In VI.30 this derivation is embedded into the general procedure for solving quadratic problems, VI.29. This last proposition explains how to find the point D (here F) under the circumstances of II.6. Essentially, it is done as above with the help of I.47 or, in more general situations, its generalization, VI.25.

We are still left with the question why the problem of II.11 should be interesting at all. A short glance into Book XIII, Prop. 8 tells us what is going on. XIII.8 is about the regular pentagon with side x and diagonal b (in our present notation). It is proved that

$$b : x = x : (b - x),$$

which is, after VI.16, equivalent to

$$b(b - x) = x^2.$$

Knowing this, we understand the essential meaning of II.11: to determine the side x of a regular pentagon when the diagonal b is given. But about this Euclid keeps quiet in Book II, and even in Book IV, where he constructs the pentagon.

THE CHAIR

This chair was once a student of Euclid.

The book of his laws lay on its seat.
The schoolhouse windows were open,
So the wind turned the pages
Whispering the glorious proofs.

The sun set over the golden roofs.
Everywhere the shadows lengthened,
But Euclid kept quiet about that.

(Charles Simic: Hotel Insomnia, New York: Harcourt, Brace 1992. With kind permission of the publisher.)

Remark

One thing should be said very clearly: Mathematical truth is not dependent on motivation. Euclid has taught this to many generations of mathematicians. Teaching a class is another thing. Students have every right to understand how particular steps are directed to a specific goal.

Propositions II.12, 13 are very important as seen from today, because they translate into the law of cosines. In the *Elements*, however, they are just stated and never again mentioned or applied. Let us first look for a motivation of the statements (which, clearly, Euclid does not supply). By the congruence theorem SAS, a triangle—and

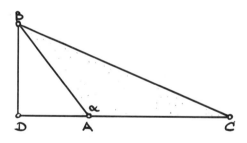

FIGURE 7.7

hence its third side—is determined by the two sides, say a, b, and the enclosed angle γ. By the theorem of Pythagoras we know the third side c if γ is a right angle. Is it possible to determine c for a general angle γ? This question is answered in II.12 for an obtuse and in II.13 for an acute angle γ. Let the first case stand for both of them.

Prop. II.12.
Let $\triangle ABC$ be an obtuse-angled triangle having $\angle BAC$ obtuse, and let BD be drawn from the point B perpendicular to CA produced [Fig. 7.7]. I say that the square on BC is greater than the squares on BA, AC together by twice the rectangle contained by CA, AD.

Proof.
By the theorem of Pythagoras I.47:

$$\square BC = \square BD + \square DC,$$
$$\square AB = \square BD + \square DA.$$

By the binomial theorem II.4,

$$\square DC = \square DA + \square AC + 2\square(DA, AC); \qquad (7.5)$$

hence

$$\square BC = \square BD + \square DA + \square AC + 2\square(DA, AC) \qquad (7.6)$$
$$= \square AB + \square AC + 2\square(DA, AC) \qquad (7.7)$$

as asserted.

Modern translation. Let $AB = c$, $BC = a$, $CA = b$, and $\angle BAC = \alpha$. Then $DA = c \cos \alpha$. Because α is obtuse, $\cos \alpha$ is negative, and we get

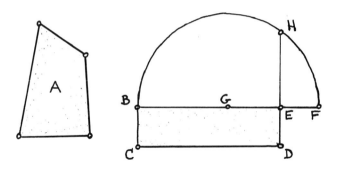

FIGURE 7.8

the result

$$a^2 = b^2 + c^2 - 2bc \cos \alpha,$$

the modern version. The sign of the cosine function will take care of both cases II.12,13. The law of cosines could develop its full power only a long time after Euclid.

Prop. II.14.
To construct a square equal to a given rectilinear figure.

Solution:
Let A be the given figure [Fig. 7.8]. Construct a rectangle $\square(BE, ED)$ of equal content with A via I.45. If this rectangle is a square, we are ready. Otherwise let BE be greater than ED. Prolong BE to F and make EF equal to ED. Bisect BF in G, describe the semicircle BHF with center G. Produce DE to H.

Assertion:
$\square(BE, ED) = \square EH.$

By II.5 for the line BF we get

$$\square(BE, EF) + \square EG = \square GF.$$

GF is equal to GH by construction, hence

$$\square(BE, EF) + \square EG = \square GH$$
$$= \square EG + \square EH \qquad \text{by I.47.}$$

Subtracting the square □*EG* on both sides and observing that *EF* = *EG* completes the proof.

This theorem is a goal for its own sake. It is the objective of a theory developed from I.35–45 via I.47 and II.5 to its culminating point II.14. It serves no other purpose than to transmit a certain knowledge or answer a question that is important in theory and, we may add, in practice as well.

8

The Origin of Mathematics 4

Squaring the Circle

....

Theorem II.14 solves an important problem: Every rectilinear figure can be squared. As usual in mathematics, a problem is solved only to beget another one. The next most prominent figure is the circle. How to square it? Proclus observes in his comment on Prop. I.45, which is the last step before II.14:

> It is my opinion that this problem is what led the ancients to attempt the squaring of the circle. For if a parallelogram can be found equal to any rectilinear figure, it is worth inquiring whether it is possible to prove that a rectilinear figure is equal to a circular area. (Proclus–Morrow 335)

The problem was indeed so prominent in classical Athens that several reports on it have survived. It must have been popular in the intellectual circles in Athens about 440–400 B.C.E. Evidence for this is a line from the comedy *The Birds* written by Aristophanes and staged on Broadway—sorry, in Athens—414 B.C.E. Aristophanes presents a certain self-appointed civic surveyor of Cloud-Cuckoo-Land.

If I lay out this curved ruler from above and insert a compass—do you see? ... by laying out I shall measure with a straight ruler, so that the circle becomes square for you. (Lines 1001–1005, after Knorr [1986])

If people didn't understand the allusion of a joke it would be bad for the success of the play. That is, "squaring of the circle" must have been understood by the general public in Athens.

The philosopher Anaxagoras ($\approx$ 500–430) is reported to have tried the quadrature of the circle during a spell in prison. At about the same time, 440–430 B.C.E., the mathematician Hippocrates of Chios had made a discovery that must have been perceived as an important step toward the solution of the problem. He had squared the first curvilinear figures, the crescent shaped lunules, or lunes. We will sketch only the most important features of Hippocrates' proof. (For details see the Notes.)

Hippocrates starts from the assertion that [the areas of] segments of a circle have to each other a ratio as the squares on their bases.

The first and simplest of his lunes are squared the following way. Starting from a right isosceles triangle (Fig. 8.1), we draw on each of its sides a semicircle. By Pythagoras's theorem and the assertion above it follows that the semicircle on the hypotenuse equals the two semicircles on the legs. If we "turn" the semicircle on the hypothenuse into the position ACB and remove the portion common to the large semicircle and the two smaller ones, we find that the isosceles triangle that remains is equal to the two lunes that arch

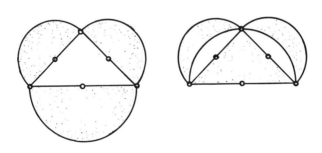

FIGURE 8.1

over its legs. Because the triangle is easily squared, we have squared the lunes.

Hippocrates treated other more complicated cases of the same type, always removing parts of circles from other parts of circles. By taking away curvilinear areas from other curvilinear areas in an artful way he in effect "cancels out" the curvature, but this can be seen only with hindsight.

Euclid, in his Prop. XII.2, proves the basic assertion of Hippocrates for circles. He employs methods of Eudoxos. If we have two circles with areas A_1, A_2 and radii r_1, r_2, then, in modern words, XII.2 says that

$$A_1 : A_2 = r_1^2 : r_2^2.$$

Consequently,

$$A_1 : r_1^2 = A_2 : r_2^2,$$

so that the ratio of the area A of a circle to the square r^2 of its radius is constant. This constant is nowadays called π, and we write

$$A : r^2 = \pi, \qquad \text{or} \qquad A = \pi r^2.$$

The second (and secondary) question is the determination of the constant π as precisely as possible. In his *Measurement of a Circle* Archimedes established

$$3\frac{10}{71} < \pi < 3\frac{10}{70}, \text{ or approximately}$$

$$3,1408 < \pi < 3.1429.$$

In the first theorem of the same paper Archimedes proves the relation between the circumference and the area of a circle:

> The area of a circle is equal to that of a right triangle in which one leg is equal to the radius and the other to the circumference of the circle.

This connects the constant π with the circumference C of the circle:

$$A = \frac{1}{2}Cr, \text{ or, using} \qquad A = \pi r^2, \qquad C = 2r\pi.$$

At the very end of antiquity, the philosophical commentator Ammonius, a disciple of Proclus, writes in a few melancholy words about what happened in the roughly 900 intervening years that separate him from Hippocrates:

> The geometers, on constructing the square equal to the given rectilinear figure, sought whether it was possible to find a square equal to the given circle. And many and the greatest [of them] sought, but did not find it. Only the divine Archimedes found an extremely good approximation, but the exact construction has not been found to this day. And this is perhaps impossible.... (Knorr [1986], 362)

A student of Ammonius in turn, Simplicius, who wrote about 540 C.E., has a very clear conception of the problem. He says:

> The reason why one still investigates the quadrature of the circle and the question as to whether there is a line equal to the circumference, despite their having remained entirely unsolved up to now, is the fact that no one has found out that these are impossible either, in contrast with the incommensurability of the diameter and the side (of the square). (Knorr [1986], 364)

Simplicius was widely read as a commentator on Aristotle in the Middle Ages. The squaring of the circle acquired the reputation of a problem that is impossible to solve. To this day it retains a proverbial character. Dante uses it in this sense in a most prominent place, within the last dozen lines of his *Divine Comedy* (written 1310 C.E.).

> Like the geometer, who deep in thought applies himself
> To measuring the circle, and cannot find,
> However much he thinks, the principle in need, ...

In the seventeenth century, when limits were studied more widely, it appeared to the mathematicians that the question of squaring the circle was really a problem about the nature of numbers, as already faintly indicated by Simplicius. James Gregory speculated in 1667 that new numbers had to be introduced in order to measure the circle. He thought of numbers introduced by a "new operation," namely taking limits beyond the ordinary operations of arithmetic and extracting roots. Leibniz and others were surprised to see π as the sum of simple infinite series, but only by complex analysis and Euler's famous formula $\exp(2\pi i) = 1$ was π firmly established in the mainstream of mathematics.

It took another 100 years after Gregory for the next great step toward the nature of π. In 1767 J. H. Lambert answered Simplicius's question by proving that π is irrational. He used the tangent function and continued fractions, methods that are largely forgotten today. The distinction between different types of irrational numbers is due to Euler and Legendre around 1750. Legendre conjectured that π was not the root of a polynomial equation

$$x^n + a_{n-1}x^{n-1} + \cdots + a_1 x + a_0 = 0$$

with rational coefficients a_i. The roots of such a polynomial are called algebraic numbers; all other real numbers are called transcendental because they transcend the power of algebraic methods. It was, however, in 1844 that Liouville exhibited the first concrete transcendental number. It was an artificially constructed number with no other interest than just being transcendental.

In the first half of the nineteenth century algebra had been developed so far that it was clear that any number (length of a segment) constructible by ruler and compass and starting from a unit segment had to be algebraic. The problem of squaring the circle (of radius 1, say) by means of Euclidean constructions using ruler and compass had thus turned into a question of the precise nature of the number π. F. Lindemann proved in 1882 that π is transcendental. Hence the problem of squaring the circle in its usual understanding is unsolvable.

The story has more in it for the general development of mathematics. First of all, *new concepts* were necessary for the solution of an old problem: the classification of numbers, and the understanding of the nature of constructible numbers.

Second, *new methods* had to be invented. The old tools were just not sharp enough for such a hard problem. Ch. Hermite had in 1873 shown that the number e is transcendental. His method came from the calculus. The differential equation $f' = f$ is characteristic for the exponential function. He approximated this differential equation very closely by polynomials on a finite interval and was able to show that this was too good to be true if e were algebraic. Lindemann followed and improved the methods of Hermite.

Third, a word about *historical reconstructions*. If one looks up a proof for the irrationality of π in a recent standard book, say *Number Theory* by Hardy and Wright, one will find a rather short and accessible proof. That proof uses the methods of Hermite. It starts from the differential equation $f'' = -f$ for the sine and cosine functions and then proceeds as indicated above, but needs only a few lines because of the more restricted statement.

If somebody wanted to find out how Lambert's proof had worked from this textbook version of Hardy's, he would be led astray completely. In many instances we are in a similar situation if we try to reconstruct from Euclid's *Elements* what had happened before Euclid. We do it nevertheless.

Squaring the Circle Again

There are still many people who do not know about Lindemann or just don't believe him. In most cases, they turn up with approximate constructions for π. Here is a particularly good one.

Start with a circle of diameter 2 and center C (Fig. 8.2). Draw the diameter AB and the tangent t at the point B. Select D on t such that $\alpha = \angle DCB$ has measure $30°$. (This can be constructed!) Make DE of length 3. Then AE will be of length π. (My pocket calculator says 3.1415333 instead of $\pi = 3.1415926$. Starting with a unit length of 1 m the difference is less than $\frac{1}{10}$ mm. This is impossible to notice in a pencil drawing.)

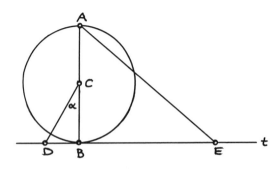

FIGURE 8.2

9

Euclid Book III

About the Circle

9.1 The Overall Composition of Book III

Definitions 1–11	
1	How to find the center of a circle
2–15	A: Chords in circles, circles intersecting or touching each other
16–19	B: Tangents
20–22	C_1: Angles in segments of circles and quadrilaterals in circles
23–29	C_2: Chords, arcs, and angles
30	How to bisect an arc
31–34	C_3: More about angles in circles
35–37	D: Intersecting chords, secants, and tangents

9.2   The Definitions of Book III

Def. 1.
Equal circles are those the diameters of which are equal, or the radii of which are equal.

In the first two Books, Euclid has used the notion of equality for rectilinear figures in the sense of "of equal content." For circles, now, "equal" has a different meaning. (For a more detailed discussion of this and the other definitions, see the Notes.)

Def. 2
[of a tangent] *A straight line is said to touch a circle which, meeting the circle and being produced, does not cut the circle.*

Def. 6.
A segment of a circle is the figure contained by a straight line and an arc of a circle.

Def. 7.
*An **angle of a segment** is that contained by a straight line and an arc of a circle. [Fig. 9.1(a)].*

Def. 8.
*An **angle in a segment** [Fig. 9.1(b)] is the angle contained by the connected straight lines when some point is taken on the arc of the segment and straight lines are connected from it to the extremities of the straight line which is the base of the segment.*

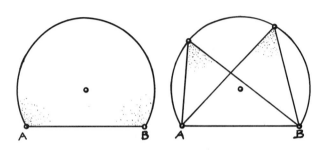

FIGURE 9.1   (a) angle *of* a segment (b) angle *in* a segment

In Def. 8 Euclid seems to presuppose the validity of Prop. III.21 about the equality of all the angles *in* a segment.

Def. 11.
Similar segments of circles are those which admit equal angles or in which the angles are equal to one another.

Here, as in Def. 7, mixed angles are used. Moreover, as in Def. 8, the equality of the angles α, α' *of* a segment is tacitly assumed. Third, the notion of similarity occurs for the first time. Equiangular triangles are treated likewise by Euclid in Props. IV. 2, 3 without using the word "similar".

In the discussion of Prop. I.5 we have seen how Aristotle used the (mixed) angles of a segment. It seems that Euclid used some older material in Book III that he did not rework thoroughly.

9.3   Book III, Part A: Chords in Circles, Circles Intersecting or Touching Each Other

Before starting with part A, Euclid poses the curious-looking problem III.1: how to find the center of a circle. The existence of a center is guaranteed by the very definition of a circle in I. Defs. 15, 16. But Euclid often enough starts from a circle and then finds its midpoint. One can only speculate that this is a remnant of an old drawing practice. Drawing a circle with a pair of compasses, which were in use in antiquity, automatically produces the center. But if an object fabricated on, say, a potters wheel, like a plate or a cup, was used to draw a circle in the sand, then indeed the midpoint had to be constructed afterwards.

The material of part A is not very coherent logically. We quote two typical examples.

Prop. III.4.
If in a circle two chords cut one another which are not through the center, they do not bisect each other.

FIGURE 9.2   Ritterstiftskirche in Wimpfen, Germany, 1280

Prop. III.11, 12.
If two circles touch each other (internally or externally) and their centers are taken, the straight line joining their centers, when produced, will fall on the point of contact of the circles.

Props. III.11, 12 have found innumerably many applications in the art of the Middle Ages. They are basic for the construction of Gothic traceries (see Fig. 9.2).

9.4   Book III, Part B: Tangents

The four propositions of this section are fundamental for elementary geometry. They transform the intuitive notion of the tangent as a line "touching the circle" into an easily usable tool by providing a simple construction for a tangent: The tangent is orthogonal to the radius drawn from the center to the point of contact.

Prop. III.16.
The straight line drawn at right angles to the diameter of a circle from its extremity will fall outside the circle, and into the space between the straight line and the circumference another straight line cannot be interposed; further, the angle of the semicircle is greater, and the remaining angle less, than any acute rectilinear angle.

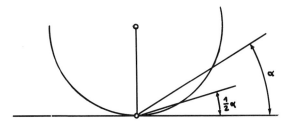

FIGURE 9.3

Before going on with tangents we need to have a look at the "hornlike" angles formed by straight lines and circles at the point of contact. Euclid states that the angle between the circle and the tangent is smaller than any rectilinear angle.

This has serious consequences for the ordering of angles according to their size. The rectilinear angle α may be bisected (and bisected again ...), but never will the resulting smaller rectilinear angle be smaller than the hornlike angle β (See Fig. 9.3). The ordering of angles, including mixed ones, is non–Archimedean in modern terms. Euclid was well aware of the situation, because on other occasions he in effect excludes non–Archimedean orderings. In Def.V.4 he speaks about magnitudes "which are capable, when multiplied, of exceeding one another," and in Prop.X.1 he has exactly the situation of III.16 and shows with the aid of Def.V.4 that the successive halves will eventually be smaller than the second magnitude. Non–Archimedean and Archimedean orderings were known some hundred years before Archimedes.

At the end of the proof of Prop. I.16 Euclid adds a corollary ("porism"):

From this it is manifest that the straight line drawn at right angles to the diameter of a circle from its extremities touches the circle.

Props. III. 18/19 are (partial) converses of the corollary.

Prop. III.17 is of special interest because of its typical use of symmetry in a mathematical argument and because of the similarity of the figure with the one of Aristotle's old proof for Prop. I.5.

Prop. III.17.
From a given point to draw a straight line touching a given circle.

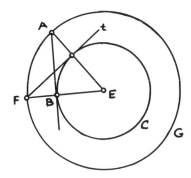

FIGURE 9.4

Let C be the given circle with center E and A be the given point (Fig. 9.4). Join A to E and let D be the intersection point of AE with the circle. Let t be the tangent to C at the point D, which is orthogonal to ED. Draw the circle G with center E and radius EA; let it intersect t in F. Find the intersection point B of FE and circle C. Claim: AB is the tangent sought. For the proof it has to be shown that AB is orthogonal to EF. The triangles $\triangle EDF$ and $\triangle EBA$ are congruent by SAS; and hence the angle at B is right because the angle at D is right. The unexpected way in which the tangent t, which at first sight has nothing to do with the problem, is used to produce the wanted tangent AB is a nice surprise and makes the construction very elegant.

9.5   Book III, Part C_1: Angles in Segments of Circles

The four most useful theorems in elementary geometry are Euclid's Prop. I.32 about the sum of the angles in a triangle, the theorem of Pythagoras I.47, the theorem about the invariance of angles in a segment Prop. III.21, and Prop. VI. 2 about the proportionality of sides in similar triangles. Prop. III.20 marks a fresh start in Book III.

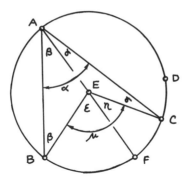

FIGURE 9.5

Prop. III.20.

In a circle the angle at the center is double the angle at the circumference when the angles have the same arc as base.

With the notation of Fig. 9.5, Euclid's proof proceeds as follows. Let $\alpha = \angle BAC$ with the arc BFC as its base. E is the center of the circle. For the first case, let E be in the interior of the angle α. The line AE is extended to F. Then $\triangle ABE$ is isosceles; hence the exterior angle ϵ is equal to 2β. Similarly, $\eta = 2\gamma$. Hence $\mu = \epsilon + \eta$ is the double of $\alpha = \beta + \gamma$. (The case of the angle δ is to be treated in the same way, with subtraction instead of the addition of angles.)

Prop. III.21.

In a circle the angles in the same segment are equal to one another.

This is an obvious corollary to Prop. III.20. It is, however, in a strict sense proved only for segments greater than half circles. We will not dwell on this point, which can be corrected easily. The next theorem is a first substantial application of Prop. 21.

Prop. III.22.

The opposite angles of quadrilaterals in circles are equal to two right angles.

Proof

Notation as in Fig. 9.6. In the triangle $\triangle ABC$, we have $\alpha + \beta + \gamma = 2R$. The angles α, α' are in the same segment defined by BC; hence $\alpha = \alpha'$ by Prop. 21. Similarly, $\gamma = \gamma'$, and because of $\delta = \alpha' + \gamma'$ we have $\beta + \delta = \beta + \gamma + \alpha = 2R$.

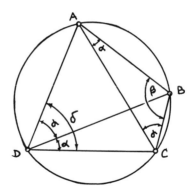

FIGURE 9.6

Doesn't this clever combination of the two invariants from I.32 and III.21 constitute a jewel of a proof? We will return to it in the following section on problems and theories.

9.6 Book III, Part C_2: Chords, Arcs, and Angles

In this section of Book III the results of the preceding one are extended in a technical way to more general situations resembling the propositions on the areas of parallelograms and triangles in Book. I.36–38. The equality of angles is transferred from "in the same segment" to "in equal segments of equal circles" and similarly for chords and arcs: Equal arcs determine equal chords, and vice versa; and the same for angles and arcs. The statements have important applications in Book IV.

9.7 Book III, Part C_3: More About Angles in Circles

In section C_1, two extreme cases have not been dealt with. The first one is the angle in a semicircle. The diameter of a circle is a straight

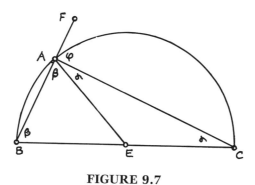

FIGURE 9.7

line, and for Euclid this does not define an angle (of 180° or 2 right angles) at the center.

Prop. III.31.

In a circle the angle in a semicircle is right.

(The proposition goes on with statements about the angles in and of segments greater or smaller than semicircles.) Prop. III.31 is ascribed to Thales by some antique sources and is called "the theorem of Thales" in Germany. (In France "the theorem of Thales" is VI.2!)

The proof again uses isosceles triangles; see Fig. 9.7. Euclid concludes that the exterior angle φ is equal to the interior angle $\beta + \gamma$; hence both must be right angles. This (and the proof of III.20 and other occasions) shows that Euclid was well aware of the later applications of theorem I.32 about the sum of the angles in a triangle and chose a formulation of I, 32 suitable to his purposes.

Prop. III.32.

If some straight line touches a circle and some straight line cutting the circle is drawn from the contact into the circle, the angles which it makes with the tangent will be equal to the angles in the alternate segment of the circle.

We will present Euclid's essential steps while abbreviating his proof and using the modern notation of Fig. 9.8. The assertion is $\alpha = \varphi$. Because of the invariance of the angle α in the segment BAC we may place A' so that BA' is a diameter of the circle. Then, by the last proposition, $\angle A'CB$ will be right; hence by I.32 we have $\alpha + \beta = R$. But because $A'B$ is a diameter and BF a tangent, also

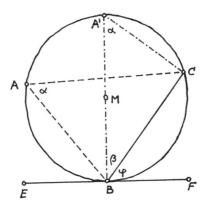

FIGURE 9.8

$\beta + \varphi = R$, from which it follows that $\varphi = \alpha$, as asserted. Notice how many previous theorems play essential roles in this one proof: I.32, III.18, 21, 31.

    The next two propositions present variants of III.32 that are useful because they create situations suitable for the application of III.21.

Prop. III.33.
To describe on a given straight line a segment of a circle admitting an angle equal to a given rectilinear angle.

Prop. III.34.
To cut off from a given circle a segment admitting an angle equal to a given rectilinear angle.

9.8   Book III, Part D: Intersecting Chords, Secants, and Tangents

The last three propositions of Book III are placed in the context of similarity geometry in today's high-school geometry, where they

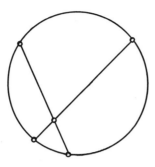

FIGURE 9.9

have easy proofs. At this stage, Euclid cannot use proportions, and he supplies complicated proofs. This is an important feature of the architecture of the *Elements*.

Implicit in III. 35/36 is the construction of an invariant, which is today called the power of a point with respect to a circle, and which has played an important role in the history of geometry. Because Prop. 35 is not used in the later books, we present the two proofs for Prop. 36/37 only.

Prop. III.35.

If in a circle two straight lines cut one another, the rectangle contained by the segments of one is equal to the rectangle contained by the segments of the other (See Fig.9.9).

Prop. III.36.

If some point (D) is taken outside a circle and from it there fall on the circle two straight lines, and if one of them cuts the circle (in A and C), the other touches it (in B), the rectangle contained by DC and DA will be equal to the square on the tangent DB.

We will restate this in a somewhat modernized notation so as to include its converse III.37. Compare Fig. 9.10.

In the situation of Fig. 9.10:

$$DB \text{ is a tangent} \quad \Leftrightarrow \quad \Box DB = \Box(DC, DA).$$

(i) Let *DB* be a tangent.

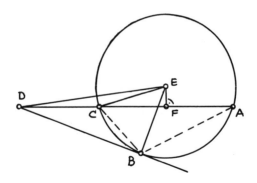

FIGURE 9.10

Similarity proof:

Consider the triangles $\triangle DBC$ and $\triangle DBA$. They have $\angle ADB$ in common. If DB is a tangent, then $\angle DBC = \angle BAC$ by III.32. By I.32 the remaining angles are equal and the triangles $\triangle DBC$ and $\triangle DAB$ are equiangular.

Using VI.4 we get

$$DC : DB = DB : DA,$$

and by VI.16,

$$\square(DC, DA) = \square DB.$$

Euclid's proof.

Euclid first considers the case where DA passes through the center of the circles. We ignore this because the methods are the same as in the other case, where the center E of the circle is not on DA. Let EF be orthogonal to DA; join E to D and B.

On the line DA, we have the situation of II.6, resulting in

$$\square(AD, DC) + \square CF = \square FD.$$

Now three applications of Pythagoras's theorem will help. By construction,

$$\square EC = \square CF + \square FE,$$
$$\square ED = \square DF + \square FE.$$

Hence

$$\square(AD, DC) + \square EC = \square ED,$$

and because $EC = EB$,

$$\square(AD, DC) + \square EB = \square ED.$$

Now, because DB is a tangent,

$$\square ED = \square DB + \square EB.$$

Substituting and subtracting $\square EB$ gives

$$\square(AC, DC) = \square DB.$$

At first sight, Euclid's proof looks much more complicated than the similarity proof. The latter, however, uses VI.4, 16, which need conceptually difficult foundations in Book V. Altogether, Euclid's proof is still simpler, in spite of being technically complicated. However, it does not present itself in a natural way. Seen from today, it is an artificial product. As long as the concepts of similarity and proportion were used in a naive way, the first proof was obvious.

(ii) The converse direction of the proof (Prop. 37) uses part (i). It is of no particular interest for us.

10

CHAPTER

The Origin of Mathematics 5

Problems and Theories

....

In section C of Book III Euclid presents the prototype of a mathematical theory. He has a clear sense of its architecture. Let us recapitulate the main steps:

> III.20 is the preparing lemma with specific information about the size of angles.
> III.21 is the main theorem about an invariant.
> III.23–29 are technical expansions of the main theorem.
> III.31/32 deal with two remaining extreme cases.
> III.33/34 modifies the extreme case 32 to get a tool for applications.

The statement of Prop. III.31 about the angle in a semicircle mentions mixed angles, and the proposition itself is connected to the name of Thales. This suggests a primary role for III.31. If we accept this and take the special case as the beginning of the theory, the picture will look even nicer: First an important special case, then the generalization III. 20/21 to the main theorem.

As a sort of windfall profit, Euclid puts in Theorem III.22 about quadrilaterals in circles. Other applications are not mentioned; the reader has to trust the author that he will see them in time. The development of the theory dominates the picture.

In contrast to this, let us outline a different approach that starts with a question about a figure. Mathematicians before Euclid certainly knew how to circumscribe a circle about a triangle as in Prop. IV. 5. From this it is evident that not every quadrangle has a circumcircle. (Three vertices of the quadrangle determine a circle, the fourth one may be on this circle or not.) The answer to the problem is obtained by generalizing the construction for a triangle: A circumcircle exists if and only if the perpendicular bisectors of the sides meet in one point. Trivial as it may look, this solution symmetrizes the problem and takes us further.

Let *ABCD* be a quadrangle in a circle. Connect the vertices to the center in order to obtain four isosceles triangles with the perpendicular bisectors from above as axes of symmetry. The sum of all the angles in a quadrangle is 4R. Exploiting the symmetries (and using the notation from Fig. 10.1), we get

$$2\alpha + 2\beta + 2\gamma + 2\delta = 4R,$$
$$(\alpha + \delta) + (\beta + \gamma) = 2R.$$

Opposite angles add up to 2R, that is Prop. III.22. The proof of the converse is easy. This gives us:

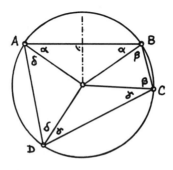

FIGURE 10.1

Theorem.

A (convex) quadrangle has a circumcircle if and only if opposite angles add up to 2R.

Corollary: Prop. III.21.

Proof:

Fix the points B, C, D and move A. The angle $\angle BCD$ remains constant and adds up with $\angle DAB$ to 2R, hence the latter angle remains constant, too.

The lesson of this story: A concrete problem could be either the start or an application of a theory. For a beginner, starting with a problem will look more natural. But the adult mathematician relishes a nice theory, and Euclid was an adult mathematician. Moreover, his proof of III.22 for the quadrangle is so elegant that he may have preferred it just for aesthetic reasons. (So do I.)

Prop. III.21 and our theorem are in effect two aspects of the same subject matter; they are logically equivalent. The theorem (or III. 22) reveals a hidden property of a figure; III.21 is a versatile tool for later applications.

More About Quadrangles

Observe that Prop. III.35 about intersecting chords (together with its converse) is a second characterization of cyclic quadrangles. Both characterizations—the theorem above and III.35—cannot simply be read off from a diagram and then verified. They are accessible only by means of proofs.

The characterization by opposite angles is not a mere curiosity of elementary geometry, but alive and well in modern mathematics. With the vertices taken as points in the complex plane $\mathbb{C}$ (and using oriented angles) it is easily transformed into the statement; Four points in the complex plane are concyclic if and only if their cross ratio is real. In this version, it is basic for the theory of the groups $PGL(2, \mathbb{C})$ and $PSL(2, \mathbb{R})$, which dominate modern hyperbolic geometry.

11
CHAPTER

Euclid Book IV

Regular Polygons

....

11.1   The Contents of Book IV

We will use the standard term "regular polygon" (or n–gon) for what Euclid calls in particular cases an "equilateral and equiangular polygon." Convexity is always tacitly assumed. Book IV follows a tight plan and has none of the subdivisions of some other books. We repeat what has been said in the section about the contents of the *Elements*. Four problems are treated systematically:

(i) to inscribe a rectilinear figure in or (ii) to circumscribe it about a given circle;

(ii) to inscribe a circle in or (iv) circumscribe it about a given rectilinear figure.

These problems are solved for

(a) triangles in general (IV. 2–5);

(b) squares (regular quadrangles) (IV. 6–9);

(c) regular pentagons (IV. 10–14);

(d) regular hexagons (IV. 15);

(e) regular 15-gons (IV. 16).

The rigorous and systematic plan of Book IV suggests that it may have been a monograph written by a single author, which Euclid incorporated as a whole in the *Elements*. The neglect of the parallel postulate and the sometimes archaic language point in the same direction. Two scholia added later to the book say that its theorems are the discovery of the Pythagoreans. Moreover, Book IV is a sort of end point in the composition of the *Elements*. It uses much of the material of the preceding books, but its contents are not really used subsequently. It is especially noteworthy that in Book XIII Euclid implicitly constructs the pentagon a second time in connection with the regular dodecahedron. On the surface there are no obvious seams in Book IV, but Euclid does appear to have deviated from the original monograph in two ways. First, he has reworked the construction of the regular pentagon to eliminate the use of proportions. And second, he does not use the first two definitions of the book about the inscription of polygons into polygons, which he presumably preserves from the original monograph.

Mathematically, the most substantial achievement of Book IV is the construction of the regular pentagon. The other theorems are comparatively easy, so one might say that the construction of the pentagon gives the whole book its raison d'être. The theory of Book IV is interesting only because of this one difficult case, which we will treat separately in the next section. Proposition IV. 2 has an especially beautiful proof, which we will present in the section about beauty in mathematics. One sample theorem will bring about the flavor of the rest of Book IV.

Prop. IV. 15.
In a given circle to inscribe an equilateral and equiangular hexagon.

Construction
(See Fig. 11.1; we abbreviate Euclid's arguments a little.) Let the given circle be $\mathcal{K}$ and its center be G. Draw the diameter AD and the (auxiliary) circle $\mathcal{H}$ with center D and radius DG. It intersects $\mathcal{K}$ in the points C and E. Prolong CG to F and EG to B.

(a) The hexagon $ABCDEF$ is equilateral. By construction, the triangles $\triangle EGD$ and $\triangle DGC$ are equilateral, their respective angles at G are one-third of $2R$. Because BE is straight, the remaining angle $\angle BGC$ is also one-third of $2R$. The three other

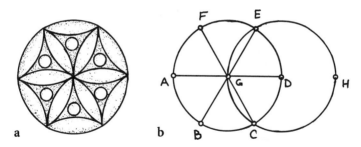

FIGURE 11.1   (a) golden disk from Mykene (1200 B.C.E.) (b) construction
of the regular hexagon

remaining angles at *G* are vertical to the previous ones; hence
all of them are equal. But equal angles stand on equal arcs
(III.26) and are subtended by equal chords (III.29). Hence the
hexagon is equilateral.

(b) It is also equiangular. The arc *BCDEF* is equal to the arc
CDEFA; hence by III.27; ∠*FAB* is equal to ∠*ABC*, and so on.

The Construction of the Incircle

In contrast to the problem of inscribing a regular *n*–gon in a given
circle, which has to be solved for each *n* individually, the construction of an incircle in a regular *n*–gon has a general solution, as does
the construction of a circumcircle. The procedure of IV.13, where
Euclid inscribes a circle in a regular pentagon, can be generalized,
since the proof of IV.13 does not make any real use of the fact that
the regular *n*–gon being dealt with is a pentagon. Euclid knows this,
because he says in VI.15 for the hexagon and in IV.16 for the 15–
gon that the construction can be carried out as in the case of the
pentagon.

11.2   The Regular Pentagon

In this section we will first study Euclid's construction of the pentagon, which in itself is a beautiful piece of mathematics, and in
the next we speculate about possible earlier constructions and the

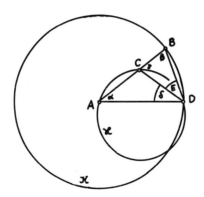

FIGURE 11.2

prehistory of Book IV. As usual, Euclid does not speak about motivations or aims of particular intermediate items (like IV, 10), but just presents the material. The decisive step towards the pentagon is the "lemma" IV. 10. The arguments in the construction itself in IV. 11 are very much like the ones in the case of the hexagon.

Prop. IV. 10.

To construct an isosceles triangle having each of the angles at the base double of the remaining one.

Construction

(See Fig. 11.2, we abbreviate) Let AB be any given line and let it be cut in C so that

$$\square(AB, BC) = \square AC, \text{ by II.11}$$

Let $\mathcal{K}$ be the circle with center A and radius AB. Make BD equal to AC and let AD, DC be joined. $\mathcal{H}$ is the circumcircle about $\triangle ADC$.

Claim: $\triangle ADB$ is as wanted.

Proof: Point B is outside of $\mathcal{H}$, the line BA cuts it in A and C, and BD "falls on it." By construction,

$$\square(AB, BC) = \square AC = \square BD.$$

Now, the decisive step of the whole argument is the application of III. 37:

$$\square(AB, BC) = \square BD \quad \Rightarrow \quad BD \text{ is tangent to } \mathcal{H}.$$

From this we get $\epsilon = \alpha$ by III.32 because α is the angle in the alternate segment of circle $\mathcal{H}$. The rest is a firework of isosceles triangles. Because $\triangle BAD$ is isosceles,

$$\beta = \epsilon + \delta = \alpha + \delta.$$

The angle γ is exterior to triangle $\triangle DAC$:

$$\gamma = \alpha + \delta = \beta.$$

Therefore, $\triangle DCB$ has two angles equal, it is isosceles, and because of construction,

$$CD = BD = AC.$$

Now $\triangle DCA$ is isosceles, resulting in

$$\alpha = \delta.$$

Hence the angles at the base, $\beta = \alpha + \delta = \alpha + \alpha$, are double the remaining one, α, as wanted.

Prop. IV. 11.

In a given circle to inscribe an equilateral and equiangular pentagon.

Construction:

Into the given circle (by the method of IV. 2) inscribe a triangle $\triangle ACD$ equiangular with the one constructed in IV. 10 (See Fig. 11.3.).

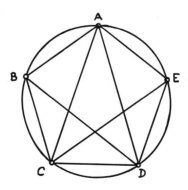

FIGURE 11.3

Bisect the angles at C and D and extend the bisectors to B and E. Join AB, BC, DE, EA.

Claim (a) The pentagon is equilateral. By construction, the five angles $\angle DAC$, $\angle ACE$, $\angle ECD$, $\angle CDB$, $\angle BDA$, are equal to each other and hence stand on equal arcs (III.26) and on equal chords (III.29). Therefore, the pentagon is equilateral. (b) By an argument similar to the one for the hexagon it follows that the pentagon is equiangular.

11.3 Speculations About the Pentagon

Apart from Book IV there is another substantial section of the *Elements* related to the pentagon in Book XIII. 1–11. For the moment we will restrict ourselves to Book IV and will return to the subject (and its relations to architecture) later. The highly polished construction of the pentagon that Euclid presents is unlikely to be historically the first and original one.

In the section about Pythagoras the interest of the Pythagoreans in the pentagram has been pointed out. Given their mathematical inclinations, an attempt by the Pythagoreans to construct the pentagon/pentagram seems very likely. Moreover, we know about Hippasus's occupation with the pentagon–dodecahedron and the scholion to Book IV that attributes its theorems to the Pythagoreans. We have, however, no direct historical source about earlier stages of the construction and have for a somewhat plausible reconstruction to resort to circumstantial evidence.

For our speculative reconstructions we will look at the permitted means of constructions, restricting them in several steps. Moreover, we will follow the useful procedure of "analysis and synthesis" that is described in a short addition to Props. XIII. 1–5 of the *Elements*:

> *What is analysis and what is synthesis? Analysis is the assumption of what is sought as if it were admitted and the arrival by means of its consequences at something admitted to be true.*
>
> *Synthesis is an assumption of that which is admitted and the arrival by means of its consequences at the end or of what is sought.*

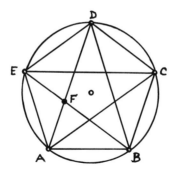

FIGURE 11.4

In fact, this is an everyday feature of mathematical research. Assume you have solved a given problem and try to figure out a way back to known statements, then try, if possible, to reverse the logical sequence. Only the second step, the synthesis, establishes a valid proof.

First Analysis of the Pentagon

Assume that a regular pentagon together with its diagonals is given (Fig. 11.4). By inspection, or by symmetry, the diagonal AD is parallel to side BC, and so on. The quadrangle $BCDF$ is a parallelogram; hence BF is equal to the side, and the points C, F, A will lie on a circle $\mathcal{K}$ with center B and radius BA. By symmetry, vertex D will be on the perpendicular bisector m of side AB. As above, DF is equal to AB.

First Synthesis with a Marked Ruler

The use of a marked ruler, or verging lines, or *neusis* in Greek, is not to be found in the *Elements*. It was, however, practiced by other Greek mathematicians, for instance Hippocrates of Chios. Pappus reports that Apollonius of Perga wrote "two books of neuses." Pappus describes the method: Given two lines in position, to place between them a straight line given in magnitude that passes (if produced)

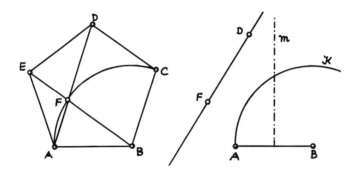

FIGURE 11.5

through a given point. We will use the first of Pappus's examples: The two lines are a semicircle and a straight line orthogonal to its base. (For an extended discussion of the marked ruler see Hartshorne, Chap. 30)

Construction: Mark the ruler with two points D, F such that DF is equal to AB (Fig. 11.5). Starting from AB construct m and $\mathcal{K}$ and then slide the ruler into position such that D is on m, F on $\mathcal{K}$, and the extended line DF passes through A. The completion of the pentagon is then easy. (But it still has to be proved that it is regular.)

It seems that at some time and for some unknown reason *neusis* constructions were decided to be unacceptable. Only ruler and compass were permitted for geometrical constructions. It is sometimes said that this goes back to Plato, but an adequate reference to Plato's writings has not been supplied.

Second Analysis of the Pentagon

If we have to forgo neusis, the difficulties with the pentagon increase dramatically. On the other hand, Euclid supplies the analysis in the context of similarity geometry, but without saying so. Starting from a given side AB of the pentagon to be constructed, we have somehow to determine the diagonal. Prop. XIII. 8 contains the relevant information. We first have to supply the definition of "extreme and mean ratio" (golden section) from Book VI, Def. 3: *A straight line is*

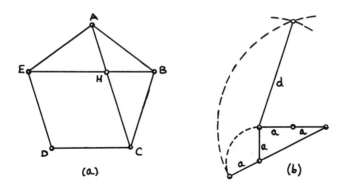

FIGURE 11.6

said to have been cut in extreme and mean ratio when, as the whole line is to the greater segment, so is the greater to the less.

Prop. XIII. 8.

If in a regular pentagon diagonals subtend to two angles in order, they cut one another in extreme and mean ratio, and their greater segments are equal to the side of the pentagon.

Proof

[abbreviated] First show that $DCHE$ is a parallelogram; hence $EH = DC = AB$ (Fig. 11.6). Next, the triangles $\triangle ABE$ and $\triangle HAB$ are equiangular. Then VI.4 implies

$$EB : BA = AB : BH.$$

But $AB = EH$; hence

$$BE : EH = EH : HB, \text{ q.e.d.}$$

By VI.16 this is equivalent to

$$\square(BE, HB) = \square EH,$$

the condition of II.11. For the moment we abbreviate $d = BE$ and f for one of the five sides of the pentagon. HB will then be $d - f$, resulting in the equation

$$d(d - f) = f^2,$$

where f is given and d is wanted. The following construction will be a little simpler with $f = 2a$. Then we have

$$d^2 - 2ad = 4a^2,$$
$$d^2 - 2ad + a^2 = 5a^2,$$
$$(d - a)^2 = 5a^2.$$

From this we can construct d easily and obtain the characteristic triangle of IV. 10 as in Fig. 11.6(b).

Second Synthesis of the Pentagon

Starting from $f = 2a$ as in Fig. 11.6(a) we obtain d and the triangle of IV, 10 with its characteristic property and may proceed as in IV. 10/11, but all the way using similarity geometry. This, for instance, simplifies the proof of III. 36/37.

Third Synthesis of the Pentagon

Not only neusis, but all similarity arguments are forbidden. This is the synthesis presented by Euclid.

Conclusion

From all this we may imagine the following stages of the origin of Book IV. First, somebody may have constructed the pentagon using neusis. Secondly, neusis was eliminated by some argument like the one in Fig. 11.6(b) or a little more elaborate using the application of areas as described in the comment on II.11. According to Eudemus, the application of areas is an ancient discovery "of the Pythagorean muse" (Proclus–Morrow 332). Hence there is good reason to believe that the construction of the pentagon (using proportions) with ruler and compass is due to Pythagorean mathematicians.

For the third stage in the history of Book IV, I assume that somebody wrote a systematic monograph, still using some type of proportion theory, but essentially giving us the text we find in the *Elements*. Such a treatise was probably written on the basis of some "Elements" containing the material of Book III rather than being incorporated into such a text. I speculate that the writing of a systematic text on regular polygons could have been motivated by Theaetetus's theory of the regular solids, so that the third stage could be located in the context of the Platonic Academy. The fourth stage of our story would be writing the new proof of IV. 10 not using proportionality, and the incorporation of the book into some 'Elements Without Proportions,' from which it passed into Euclid's *Elements*.

12

The Origin of Mathematics 6

The Birth of Rigor

....

Our historical reconstructions about the pentagon may be hypothetical. Nevertheless, we can use them as an example for some remarks on rigor in mathematics. What is meant by saying that an argument is rigorous and not just intuitively right?

For the square and the regular hexagon we have such easy constructions that we need no special theory for justifying them. The same is true for the construction of the regular pentagon using a marked ruler. But by restricting the permitted tools, the intuitive basis is lost. It is still possible to find the way to the division in extreme and mean ratio, but after that, only theory can help. We need the theorem of Pythagoras and the application of areas in order to find access to the division in extreme and mean ratio. This is what makes the construction of the pentagon a problem so much deeper than the construction of the hexagon.

When the tools are even further restricted by forbidding ratio and proportion, the situation looks almost artificially complicated. But in reality, this is not so, as our discussion of III.36 has shown. The arguments are much more clearly arranged if we use only a few principles. If we know the general view and can justify every single step, then our argument is rigorous, but this means that everything is fitted into a closed theory. Once this theory is established, it takes hold of similar or easier problems, which had previously been treated on an intuitive basis, like the square and the hexagon. Even the elementary constructions of triangles, Prop. IV. 2–5, are now fitted into the system. (Compare our discussion of Prop. I.5.)

There remain questions about the justification of every single step. Part C of Book III does this for the angles in circles, of which heavy use is made in Book IV. On the other hand, there are problems at the very beginning of Book I like the proof of the congruence theorem, Prop. I.4, or some of the foundational questions in Book III. The question is, At what stage will one accept a theory as complete? I, for one, regard Euclid's Book IV as a rigorous one—even if he neglects his own parallel axiom in IV. 5.

FIGURE 12.1 Medieval mathematician (Bremen Cathedral, east crypt. Exact date unknown)

What has been said so far is about geometric constructions, where pictures may or may not help. The situation changes completely with the introduction of abstract concepts that have no intuitive counterpart. Historically, the first and most significant one of these is the concept of incommensurability, or irrationality. There is no intuitive background that would allow one to control the validity of assertions about such abstract concepts. This forces the investigator to use precise definitions and logical deductions in order to arrive at valid results. The most prominent example of this is the theory of proportions in Book V.

A remark should be added about the justification of every single step that is of a more pragmatic nature. On the one hand, one may work out so many details that one cannot see the forest for the trees. A proof for the irrationality of $\sqrt{2}$ extending over two dozen pages in a heavy logical formalism is an impenetrable jungle. On the other hand, a barely sufficient sketch of arguments may be enough for the expert, but incomprehensible to a more general audience. The ordinary mathematician will take logic for granted in his proofs. Euclid did the same with ordering. In the end, what is a rigorous presentation depends upon the level of sophistication of the audience. Nobody would tolerate outright mistakes or deceptions, but if we insist on a "closed theory," there is no way around axioms and deductions, with the details of the exposition suitable to the audience.

13

C H A P T E R

The Origin of Mathematics 7

Polygons After Euclid

··

13.1 What We Missed in Book IV

In Prop. IV. 16 Euclid constructs a regular 15–gon by superimposing an equilateral triangle on a regular pentagon (Fig. 13.1).

Implicit in this solution is a general principle: If we are able to construct the regular r– and s–gon, and moreover, we know integers x, y such that $xr + ys = 1$, then we can construct the rs–gon as well. We need an arc of $\frac{1}{rs}$ of the full circle and x, y as above:

$$\frac{1}{rs} = \frac{xr + ys}{rs} = x\frac{1}{s} + y\frac{1}{r}.$$

Hence this combination gives us the desired arc. In the case of the 15-gon we had

$$\frac{2}{5} - \frac{1}{3} = \frac{1}{15}.$$

Since for any integers r, s with greatest common divisor $gcd(r, s) = 1$, we can use the Euclidean algorithm (VII.1,2) to find

113

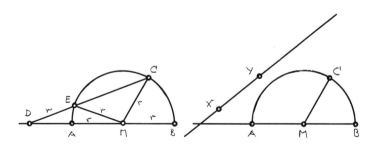

FIGURE 13.1

x, y with the desired properties, we are easily led to item (iii) listed below. This general formulation is, of course, modern. Euclid has no reason to formulate the general principle (iii) because he needs only the specific case he deals with. He generally refrains from hypothetical statements like the one above: "If $gcd(r, s) = 1$, then ..." if he does not have r, s with the desired property.

13.2 What Euclid Knew

With the above reservations we may say that Euclid (in principle) "knew" item (iii) below, and certainly he (or the pre–Euclidean author of Book IV) knew the trivial observations (i) and (iv):

(i) For all $n > 1$, the 2^n-gon is constructible.

(ii) The 3– and the 5–gons are constructible.

(iii) If the r– and s-gons are constructible and $gcd(r, s) = 1$, then the $r \cdot s$-gon is constructible.

(iv) If the n-gon is constructible and k divides n, then the k–gon is constructible.

Given (i)–(iv), the problem for general n is reduced to prime powers p^i for odd primes p.

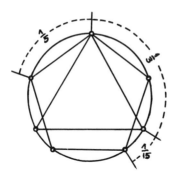

FIGURE 13.2

13.3 What Archimedes Did

By a certain variant of the *neusis* procedure, Archimedes was able to construct the regular 7–gon. By the ordinary *neusis*, he could construct the 9-gon. In fact, this *neusis* shows how to trisect an angle. (Knorr [1986], 185).

Look at Fig. 13.2 and proceed by analysis and synthesis. Let the circle with center M and radius r and the line DEC with distance $CE = r$ be given.

Then we find, for the angles α and β, using I.32 and I.5:

$$\alpha = \angle MDE + \angle MCD$$
$$= \angle MDE + \angle MEC$$
$$= \angle MDE + 2\angle MDE$$
$$= 3\beta.$$

Synthesis. Let angle $\alpha < 90°$ be given. Mark distance $XY = r$ on the ruler and slide it into position such that X is on the (produced) line AB, Y is on the circle, and the (produced) line XY passes through C. The resulting angle β will be $\frac{1}{3}\alpha$.

Application: Trisect an angle α of 60° to find $\beta = 20°$. With 2β at the center of a circle construct the regular 9–gon.

13.4   What Gauss Proved

Carl Friedrich Gauss (30 April 1777–23 Feb. 1855) started his scientific diary on 30 March 1796 with the following entry:

> Foundations on which the division of the circle rests, that is its geometrical divisibility in seventeen parts, and so on.

As this short note reveals, the teenager Gauss had not only found the construction of the 17–gon, but also the general principles behind it. He explicitly states in his first publication, in April 1796, that the 17–gon is only a special case of his investigations. He had already begun working on his great work *Arithmetical Investigations (Disquisitiones arithmeticae)*, which appeared in 1801 and in which he proved the general theorem (article 365):

> For an odd prime number p, the regular p^i-gon is constructible by ruler and compass if and only if $i = 1$ and p is a prime of the form $p = 2^{2^k} + 1$, i.e., if p is a so-called Fermat prime.

The reduction of the problem to prime powers as above is done by Gauss in art. 336.

Not all Fermat numbers $F_k = 2^{2^k} + 1$ are prime. Except for the first few values $F_0 = 3$, $F_1 = 5$, $F_2 = 17$, $F_3 = 257$, and $F_4 = 65537$, which are prime, all others with $k < 24$ are known to be composite. As long as no general result about the Fermat numbers is obtained, the question about the constructibility of regular polygons remains open.

Note. We have seen how to construct a regular 9–gon with a marked ruler. Gauss's theorem shows that the 9-gon is not constructible by ruler and compass. Hence the use of a marked ruler is definitely a stronger method than the ordinary ones in geometry. (For more and precise information, see Hartshorne, section 30.)

13.5 How Gauss Did It

For this part we assume that the reader is familiar with the elementary properties of the complex numbers. This new tool stands the test by solving an old problem. We will present Gauss's method only for the pentagon and in a somewhat modernized version. For the 17–gon and a complete treatment, see Hartshorne, chapter 6. In the complex plane, the regular n–gon is represented by the nth roots of unity, that is, the solutions of the equation

$$z^n - 1 = 0.$$

These solutions are

$$\zeta = \cos\frac{2\pi}{n} + i\sin\frac{2\pi}{n}$$

and its powers $\zeta^2, \ldots, \zeta^{n-1}, \zeta^n = \zeta^0 = 1$. We will investigate only the case $n = 5$, that is, the regular pentagon, from a new vantage point (Fig. 13.3).

The representation

$$\zeta = \cos\frac{2\pi}{5} + i\sin\frac{2\pi}{5}$$

will not help us. We have to use the algebraic equation

$$0 = z^5 - 1 = (z - 1)(z^4 + z^3 + z^2 + z + 1).$$

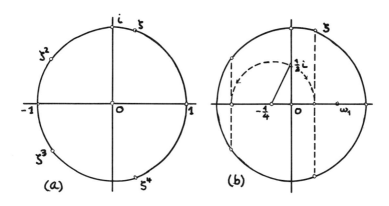

FIGURE 13.3

The first factor represents the solution $\zeta^0 = 1$ and is of no further interest. It just fixes the position of the pentagon in the circle. Because all our solutions will be different from zero, we may divide the second factor by z^2 and obtain the new equation

$$
\begin{aligned}
0 &= z^2 + z + 1 + z^{-1} + z^{-2} \\
&= z^2 + 2 + z^{-2} + z + z^{-1} - 1 \\
&= (z + z^{-1})^2 + (z + z^{-1}) - 1.
\end{aligned}
$$

With this little trick we have obtained a quadratic equation for $w = z + z^{-1}$:

(∗)
$$
\begin{aligned}
w^2 + w - 1 &= 0, \\
\left(w + \frac{1}{2}\right)^2 &= \frac{5}{4}, \\
w_{1,2} + \frac{1}{2} &= \pm \frac{\sqrt{5}}{2}.
\end{aligned}
$$

For a geometric interpretation of this, we have to observe that

$$
\zeta^{-1} = \zeta^4 = \overline{\zeta} \quad \text{and} \quad \zeta^{-2} = \zeta^3 = \overline{\zeta^2}.
$$

Our two solutions for w then amount to

$$
\begin{aligned}
w_1 &= \zeta + \zeta^{-1} = 2 \cdot \text{ real part of } \zeta, \\
w_2 &= \zeta^2 + \zeta^{-2} = 2 \cdot \text{ real part of } \zeta^2.
\end{aligned}
$$

This gives us something constructible:

$$
\begin{aligned}
\text{real part of } \zeta &= -\frac{1}{4} + \frac{\sqrt{5}}{4}, \\
\text{real part of } \zeta^2 &= -\frac{1}{4} - \frac{\sqrt{5}}{4}.
\end{aligned}
$$

Figure 13.3 (b) shows how it is done by using the right triangle with vertices 0, $\frac{1}{2}i$, $-\frac{1}{4}$. For the final determination of $\zeta = x + iy$ we know x and have to solve a second quadratic equation,

$$
x^2 + y^2 = 1
$$

for y. The result is

$$y = \frac{1}{4}\sqrt{2}\sqrt{5 + \sqrt{5}}, \quad \text{for } \zeta,$$

$$y = \frac{1}{4}\sqrt{2}\sqrt{5 - \sqrt{5}} \quad \text{for } \zeta^2.$$

This gives us

$$f = \frac{1}{2}\sqrt{10 - 2\sqrt{5}}$$

for the side f of a regular pentagon inscribed in the unit circle and $d = \frac{1}{2}\sqrt{10 + 2\sqrt{5}}$ for its diagonal.

13.6 The Moral of the Story

Note (i). Our quadratic equation ($*$) for w is similar to the one we found for the diagonal of a pentagon with side $f = 1$. There we had

$$d^2 - d = 1.$$

If we keep $f = 1$ and let $d = f + w = 1 + w$, the geometric equation becomes

$$(w + 1)^2 - (w + 1) = 1, \tag{13.1}$$
$$w^2 + w = 1, \tag{13.2}$$

the same as above. There are many different constructions of the pentagon, but closer inspection will always turn up this same quadratic equation (or a close relative as above). This equation is what one calls the **abstract essence** of the problem. Abstraction makes clear what the real nucleus of the problem is, apart from all different disguises.

Note (ii). With what was called "a little trick" we reduced the solution of the biquadratic equation for ζ to the successive solution of two ordinary quadratic equations. The trick is in fact Gauss's general method: Find the roots different from 1 for the equation $z^n - 1 = 0$ by solving quadratic equations successively, which can be done by ruler and compass. This new method, based on the new tool of complex numbers, is the unifying idea for the construction of regular polygons. Euclid had to deal with each case separately. Gauss, by abstraction, found the general solution. Generalization and abstraction made the problem accessible and the solution transparent.

13.7   What Plotinus Has to Say About All This

The neo-Platonic philosopher Plotinus ($\sim$ 200–270 C.E.) wrote a treatise about beauty. He has found the right words for what we have seen in a particularly significant case:

> But where the Ideal Form has entered, it has grouped and coordinated what from a diversity of parts was to become a unity: it has rallied confusion into cooperation: it has made the sum one harmonious coherence: for the Idea is a unity and what it moulds must come to unity as far as multiplicity may.
>
> And on what has thus been compacted to unity, Beauty enthrones itself, giving itself to the parts as to the sum: when it lights on some natural unity, a thing of like parts, then it gives itself to that whole. Thus, for an illustration, there is the beauty, conferred by craftsmanship, of all a house with all its parts, and the beauty which some natural quality may give to a single stone. (Plotinus, *First Ennead* VI, "On Beauty" 2. p. 22, cf. Plotinus [1952])

14

Euclid Book V

The General Theory of Proportions

14.1 Proportions Outside of Mathematics

One of the main discoveries of Pythagoras was the relation between musical harmonics and the ratios of segments on a monochord, the simplest one being 2 : 1 for the octave. Much has been written about this. We will complement it by a short look at architecture and the arts.

Some of the geometric terminology seems to come from the building trade, where exact designs were necessary. About 540 B.C.E. the temple of Apollo in Corinth was built. It is the earliest known temple with a clear proportion, length : breadth = breadth : height. Much of the temple of Zeus in Olympia, completed in 456 B.C.E., is governed by the ratio of 1 : 2, starting from two-foot-wide tiles on the roof to a distance of 16 feet between the centers of the columns. A little later, in the years 447–432, the architects Ictinos and Callikrates built the Parthenon temple in Athens. The ratio 9 : 4, that of the smallest square numbers, determines the whole building. Again we find

$$\text{length} \ : \ \text{breadth} \ = \ \text{breadth} \ : \ \text{height}$$
$$81 \ : \ 36 \ = \ 36 \ : \ 16$$

At about the same time, Polycletus, together with Phidias, the most famous sculptor in classical Athens, wrote a treatise about the proportions of the human figure, called the "Canon." Unfortunately, it is lost, but again we see the importance of proportions outside of mathematics. (Cf. Steuben [1973]) Plato alludes to this in the Sophist (235d–236d), where he says, "The perfect example of this consists in creating a copy that conforms to the proportions of the original in all three dimensions...."

That proportions were commonplace may also be inferred from a fragment by the philosopher Zeno (about 460 B.C.E., the one with Achilles and the tortoise). He speaks about rather extreme ratios when he says:

Isn't there a ratio between a bushel of millet and a single grain of millet? In the same ratio will be the sounds made by a falling bushel of millet and a single falling grain, even by the ten-thousandth part of a grain. (Zeno, DK 29 A 29)

Aristotle holds proportions in high esteem. He even defines justice as being nothing but proportion in his *Nicomachean Ethics*. Quite different from those of modern times are his opinions about the distribution of tax money and social justice: Justice is "a distribution of moneys from the public treasury which follows the same ratio as the respective contributions (to the public revenues) bear to one another." (A person who pays no taxes should get no money from the state. Aristotle–Heath p. 272, Nic. eth. 1131b29–32.)

At the same place Aristotle mentions the more technical procedures of alternation (explained below) and composition of a ratio. (Defs. 12 and 14 of Book V.)

More than in arts and philosophy, Greek mathematics abounds with proportions. The overarching character of the general theory of proportions was expressed by Eratosthenes ($\approx$ 250–200 B.C.E.), who said that they are "the unifying bond of the mathematical sciences" (Proclus–Morrow 36). The skillful handling of proportions was a main tool in mathematics up to the time of Galileo and Newton, who used them in masterly fashion.

14.2   General Remarks About Book V

There are no very distinct sections in Book V. It has but one single subject: the theory of proportions for general magnitudes. A scholion tells us that the theorems of Book V come from Eudoxus—and we may assume that the definitions do as well. There are very detailed studies about Book V, notably by Becker, Beckmann, and Mueller. We will not be as subtle as these authors but will restrict ourselves to a general overview. First of all, we have to pay attention to three points:

(a) Book V is much more abstract in character than Euclid's other books. Its propositions apply to various kinds of magnitudes like lines, surfaces, solids, maybe even times or angles. With respect to Plato's philosophy one might say that due to the higher level of abstraction, we are closer to Plato's ideas (or forms). Plato writes (*Phaedrus* 265de):

The first is that in which we bring a dispersed plurality under a single form (idea) seeing it all together—the purpose to define so–and–so, and thus to make plain whatever may be chosen as the topic for exposition.

The definitions at the beginning of Book V do exactly what Plato postulates: They bring a dispersed plurality under one single concept.

(b) Book V is independent of the preceding books. It could have been either a monograph or the introduction to some "Elements" written in the school of Eudoxus. As it stands in Euclid's

Elements, it opens a new gate, especially to the geometry of
similarity in Book VI.

(c) The general and abstract theory cannot have been the first
treatment of proportions in Greek mathematics. Leon and Hippocrates certainly presented the main theorems about ratios
and proportions in their "Elements." There are various speculations by historians about the definitions and proofs in these
pre–Euclidean theories, but nothing is known for sure.

We will observe two quite different aspects in discussing Book V.
The first one is the abstract nature and the fine points of the definition of "in the same ratio." The second and easier one is intended
to make a modern reader familiar with the intuitive meaning of the
various propositions that are stated in an abstract manner in Book
V. We will deal with the second aspect first.

14.3   Proportions in a Modern Version

In order to make Book V more accessible to the modern reader we
will use the anachronistic but effective modern algebraic notation.
Next to it we will present a geometric interpretation for lines that is
closer to Euclid's spirit. Regard all letters as representing lengths of
lines and treat them as (positive) real numbers.

Arithmetically, we can define

$$a : b \qquad \text{by} \qquad \frac{a}{b},$$

and have

$$a : b = c : d \qquad \Leftrightarrow \qquad ad = bc.$$

Geometrically, following Euclid's usage, we leave "ratio" undefined and interpret it as something like the inclination or slope of
the diagonal in a rectangle with sides a, b. "To be in the same ratio"
should then mean "to be about the same diagonal" (as in VI.24, 26);
see Fig. 14.1.

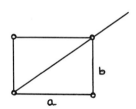 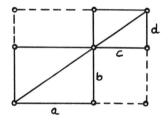

FIGURE 14.1

Now Props. I.43 and VI.16 translate into

$$a : b = c : d \quad \Leftrightarrow \quad ad = bc, \quad \text{as above.}$$

We would then, e.g., have Prop. V, 12 as a simple arithmetical or geometrical consequence:

$$a : b = c : d \quad \Rightarrow \quad (a + c) : (b + d) = a : b.$$

An important proposition with very many applications concerns *alternation* (Greek *enallax*) in a proportion:

Prop. V. 16:
$$a : b = c : d \quad \Leftrightarrow \quad a : c = b : d.$$

Arithmetical proof:

$$a : b = c : d \quad \Leftrightarrow \quad ad = bc,$$
$$a : c = b : d \quad \Leftrightarrow \quad ad = cb.$$

Because of the commutative law for the multiplication (of numbers), we have $bc = cb$, and hence the assertion. From the modern foundations of geometry one knows that the commutative law for the multiplication of the coordinates translates into the strongest geometrical axiom (i.e., the configuration theorem of Pappus). So in hindsight, it is no wonder that Prop. V. 16 has important applications. The geometric Fig. 14.2 shows that Prop. V. 16 is not obvious. Observe that one would get the same picture by placing the given segment (here b) in Prop. I. 44 in two different ways. Various other interpretations are possible, for instance switching from the triangles $\triangle SXA$ and $\triangle SUC$ to the triangles $\triangle SXY$ and $\triangle SUV$.

Another important means for the manipulation of proportions is called *ex aequali*, which means:

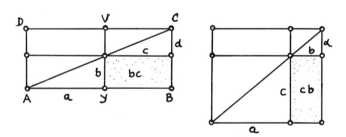

FIGURE 14.2

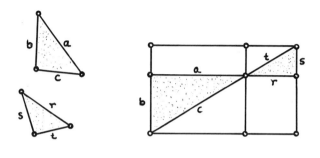

FIGURE 14.3

Prop. V. 22:

$$
\left.
\begin{array}{l}
a : b = r : s \\[1ex]
\text{and} \\[1ex]
b : c = s : t
\end{array}
\right\} \Rightarrow \quad a : c = r : t.
$$

Arithmetically, this is very simple. Just multiply and cancel. Geometrically it is more interesting. Look at two similar triangles with sides a, b, c and r, s, t. (See Fig. 14.3.) Similarity of triangles means that three pairs of corresponding angles are equal. Because the sum of the angles in a triangle is invariant, equality of two pairs of angles suffices. Similarity may also mean, by VI.4/5, that the ratios of corresponding sides are respectively equal. Because of *ex aequali*, two pairs of sides suffice.

14.4   The Definitions of Book V

Def. V. 1.

A magnitude is a **part** *of a magnitude, the less of the greater, when it measures the greater.*

Def. V. 2.

The greater is a **multiple** *of the less when it is measured by the less.*

Euclid does not say what a magnitude is, and he defines "part" or "multiple" by the undefined "measuring." In Props. V. 1–6 he elaborates the properties of "measuring" or multiplying a magnitude a by a natural number (positive integer) n so as to get na.

Def. V. 3.

A **ratio** *(logos) is a sort of relation in respect of size between two magnitudes of the same kind.*

Again "ratio," as fundamental for Euclid as "set" for a modern mathematician, remains undefined, like "set." The Greek word *logos* has a very wide range of meanings. One particularly important example of the problems in translating *logos* occurs at the beginning of the gospel of St. John: "In the beginning there was the *logos*. . . ." Goethe's Faust despairs about the translation of this sentence into his beloved German. In spite of the difficulties in philosophy and theology, the mathematical meaning of *logos* becomes clear from the context.

What it means to be "of the same kind" is apparent from the examples and is further specified in the next definition. Angles, including horn angles, are magnitudes of the same kind, as are lines, areas, and so on. Implicit in the next definition as in the "size" in Def. 3 is a (linear) ordering of the magnitudes of the same kind and the postulate that this ordering should be Archimedean. That means that infinitesimal magnitudes like horn angles are excluded. In modern terms, Def. 4 says: The ratio $a : b$ exists if there exist numbers n, m such that $na > b$ and $mb > a$.

Def. V. 4.

Magnitudes are said to **have a ratio** *to one another if they are capable, when multiplied, of exceeding one another.*

Def. V. 5.

Magnitudes are said to **be in the same ratio***, the first to the second and the third to the fourth, when, if any equimultiples whatever are taken of the first and third, and any equimultiples whatever of the second and fourth, the former equimultiples alike exceed, are alike equal to, or alike fall short of, the latter equimultiples respectively taken in corresponding order.*

Def. V. 6.

Let magnitudes which have the same ratio be called **proportional** *(analogon).*

Definition 5 is the center piece of Book V. We will first translate it into our beloved formulas. This can be done without Faust's despair. Mathematics is independent of times and languages, it is a cultural invariant of mankind.

Let a, b, c, d be the four magnitudes and m, n natural numbers. Then we have $a : b = c : d$ if for all n, m:

$$na > mb \iff nc > md,$$
$$na = mb \iff nc = md,$$
$$na < mb \iff nc < md.$$

We will go a step further and tie this to the modern theory of real numbers. This is no longer a translation but an interpretation. For this purpose, specify a, b as positive real numbers such that

$$a : b = ab^{-1} = x,$$
$$c : d = cd^{-1} = y.$$

Then we have, for the first line above,

$$na > mb \iff nc > md,$$
$$a > \frac{m}{n}b \iff c > \frac{m}{n}d,$$
$$x = ab^{-1} > \frac{m}{n} \iff y = cd^{-1} > \frac{m}{n},$$

and this for all rational numbers $\frac{m}{n} = r$.

In the language of sets this reads: The (positive) real numbers x, y are defined to be equal if the set of all rational numbers $r < x$ is equal to the set of all rational numbers $< y$. The equality of real numbers is defined via the equalitiy of sets of rational numbers. This looks very much like the Dedekind cuts of 1858/1872, but there is a fundamental difference. Euclid always starts from given magnitudes, e.g., from lines constructed in his geometrical theories. Dedekind ignores the (pre–)existence of (some) real numbers and starts from the rationals in order to *create* the real numbers. He wrote this expressively in a letter to his friend H. Weber (24 Jan. 1888):

> We are of divine stock and there is no doubt that we have creative power not only in material things (railways, telegraphs) but in particular in spiritual things. . . . I prefer to say that I create something new (different from the cut). . . . We have every right to adjudge ourselves such creative powers. . . .

Self–confident as the Greeks may have been, this is not the style of Euclid.

The definitions of Book V go on to describe properties of and manipulations of ratios. Def. 7 describes the greater relation for ratios, Def. 12 *alternate* ratios, Definition 17 a ratio *ex aequali*. We have already seen what these concepts mean in geometry.

14.5 The Propositions of Book V

The first six propositions of Book V concern multiples of magnitudes. They are preparatory for the main theorems about ratios and proportions. We translate and abbreviate them in modern formulas for magnitudes a, b and natural numbers m, n:

1. $n(a + b) = na + nb$
2. $(n + m)a = na + ma$
3. $n(ma) = (nm)a$
4. $a : b = c : d \quad \Rightarrow \quad (ra) : (sb) = (rc) : (sd)$
5. $r(a - b) = ra - rb$ (if $a > b$)
6. $(r - s)a = ra - sa$ (if $r > s$)

These statements closely resemble (some of) the modern axioms for vector spaces. The magnitudes play the role of vectors; the numbers are the scalars. This conceptual difference between numbers and magnitudes may partly explain why Euclid does not apply his abstract theory of proportions for magnitudes to numbers in Book VII. It has always been a mystery why Euclid started all over again in Book VII with a theory of proportion for numbers. The difficulties would be the same as with a one-dimensional vector space with real scalars, where the real numbers are simultaneously scalars and vectors at the same time.

After these preparations, Def. 5 is easier to work with. In Props. 7–11 ratios are treated like mathematical objects, but Euclid never uses a ratio as we do when we use a ratio of integers as a rational number.

Props. V. 7/9: $a = b \Leftrightarrow a : c = b : c$,
$$c = d \Leftrightarrow a : c = a : d.$$

Props. V. 8/10: $a > b \Leftrightarrow a : c > b : c$.

Prop. V. 11: *The equality of ratios is transitive.*

Propositions 12–25 are about the manipulation of proportions. We have seen some examples in Section 3 above. The proofs are sometimes rather intricate. In order to get the flavor of it, we will look at Proposition 17 and 19. Prop. 17 contains the technical part, and 19 contains what is wanted for later applications. We state 19 first.

Prop. V. 19:
If, as a whole is to a whole, so is part subtracted to part subtracted, the remainder will also be to the remainder as whole to whole.

Using the geometric interpretation of Section 3, we can understand this at a glance (see Fig. 14.4). (A variant of this reappears in Prop. VI.26)

Euclid even uses the language of segments in V 17/19, where he says, "Let AB be the whole magnitude and part AE be subtracted, with remainder EB, ...," and finally states the assertion

$$AB : CD = BE : DF \quad \Rightarrow \quad EB : FD = AB : CD.$$

For the proof of Prop. 19 Euclid applies Props. 11, 16 (alternation), and 17, but not Def. 5 in an explicit way. This is what he does in Prop. 17:

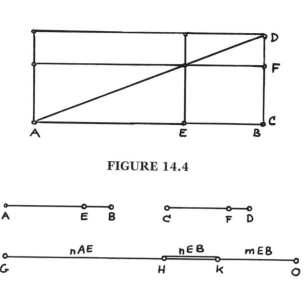

FIGURE 14.4

FIGURE 14.5

Prop. V. 17.

If magnitudes are proportional compounded, they will also be proportional separated.

Again Euclid uses the language of segments and includes a figure.

From Euclid's figure we understand that he really means the abstract situation and not our geometrical interpretation as in Fig. 14.4. In the notation of Fig. 14.5 we have $AE+EB = AB$ and $CF+FD = CD$, and the assertion is

$$AB : BE = CD : DF \quad \Rightarrow \quad AE : EB = CF : FD.$$

We will follow Euclid's proof in modern notation. We have to check Def. 5 for AE, EB, CF, DF. Definition 5 requires us to show

$(*)$ $nAE > mEB \quad \Rightarrow \quad nCF > mFD$, and the same for $<$ and $=$.

As a sort of bridge between the above magnitudes, Euclid introduces nEB and nFD.

Step 1 of the proof is to use Props. V. 1, 2 in order to show

$$nAE + nEB = nAB,$$
$$mEB + nEB = (m + n)EB,$$
$$nCF + nFD = nCD,$$
$$mFD + nFD = (m + n)FD.$$

Step 2 is to invoke the hypothesis

$$AB : BE = CD : DF,$$

and use the above multiples and Def. 5 to obtain

(**) $$nAB > (n + m)EB \Rightarrow nCD > (n + m)FD,$$

and similarly for = and <.

Step 3. For the sake of logical clarity we write (*) as $\mathcal{A} \Rightarrow \mathcal{B}$ and $nAB > (n + m)EB$ as $\mathcal{C}$.

Now Euclid starts with

$$\mathcal{C} : nAB > (n + m)EB,$$

substracts nEB on both sides, and gets

$$\mathcal{A} : nAE > mEB.$$

Step 4. Again starting from $\mathcal{C} : nAB > (n + m)EB$, by (**) Euclid has

$$nCD > (n + m)FD.$$

Now nFD is subtracted on both sides to derive

$$\mathcal{B} : nCF > mFD.$$

Step 5. We have step 3: $\mathcal{C} \Rightarrow \mathcal{A}$ and step 4: $\mathcal{C} \Rightarrow \mathcal{B}$. From this Euclid concludes that $\mathcal{A} \Rightarrow \mathcal{B}$, an obvious logical error, but not a serious one. He could have started with $\mathcal{C}$, worked backwards as in step 3 and then continued with step 4. This finishes the proof of Prop. V. 17.

There are two aspects to this proof.

First, it is rigorous. There is a definition that has to be checked, and appeal to a figure is not allowed. (The little logical error does not matter. It does not affect the principle.)

Second, it has an intuitive, or creative, component. Certainly the definition has to be checked. But how? At this point the idea of using the two bridge segments $HK = nEB$ and $MN = nFD$ leads to the solution. One might hit on this idea by some sort of analysis, but in general there is no standard procedure that will come to help. Logic and a creative component are intertwined in a good mathematical proof.

15

C H A P T E R

Euclid Book VI

Similarity Geometry

15.1 The Overall Composition of Book VI

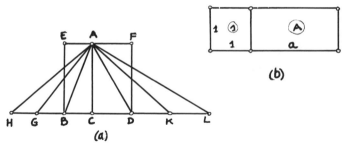

FIGURE 15.1

15.2 The Basis of Similarity Geometry

The basic theorem of Book VI looks innocent enough, but it is the foundation of Euclid's similarity geometry.

Prop. VI.1.
Triangles and parallelograms which are under the same height are to one another as their bases.

By "the triangle" Euclid means, as usual, its area. Let $\triangle_1 = \triangle ABC$ and $\triangle_2 = \triangle ACD$ be two triangles under the same height. Then the theorem says that

$$\text{Area } \triangle ABC : \text{Area } \triangle ACD = \text{Line}BC : \text{Line}CD.$$

For the proof, Def. 5 of Book V has to be verified. This is not difficult, for if the line $HC = nBC$, then the corresponding triangle $\triangle HCA$ will have area $n \triangle BCA$, and the same argument works for line CD and $\triangle CDA$ (Fig. 15.1). Hence

$$nBC > mCD \Leftrightarrow n \triangle BCA > m \triangle CDA,$$

and so on, and Prop. VI.1 is proved.

Comment

The point of this proposition is that it connects magnitudes of different kinds: areas and lengths. Let us for a moment use modern formulas: Let $a = BC$ and $b = CD$ be the bases of the triangles with common height h. Then we have

$$\triangle_1 = \frac{1}{2}ah \qquad \text{and} \qquad \triangle_2 = \frac{1}{2}bh,$$

and the consequence

$$\triangle_1 : \triangle_2 = a : b$$

is obvious.

But this involves a deception, in modern geometry as well as in Euclid's development. The formulas follow from Prop. VI.1, and not the other way round!

Why is this true? If one knows the formula ah for the area of a rectangle with sides a and h, then the formula for the triangle follows readily (cf. Prop. I.41). Hence it is enough to prove the formula for rectangles. We may even for the first step let the height $h = 1$, because for fixed a and variable h the proof will be the same "in the other direction." Now we are confronted with the basic question, What does it mean to measure a rectangle? Or even what does it mean to measure a segment AB? One fixes a unit segment, say OE, and looks for a real number a such that $AB = aOE$, or in other words,

$$AB : OE = a : 1.$$

In our discussion of Def.V.5 we have seen how this is related to the theory of real numbers.

Prop. VI.1 goes a step further. Once the unit segment OE is fixed, so is the unit square to which one assigns area (or measure) 1. Measuring a rectangle means measuring it as a multiple of the unit square. Reducing this question to measuring the sides means reducing the 2–dimensional measure to the 1–dimensional measure of segments. Seen this way, Prop. VI.1 is the first significant precursor of the construction of product measures in measure theory.

The difficulties in VI.1 arise from incommensurable segments. As usual, Euclid keeps quiet about that—up to now he hasn't even mentioned that concept! In Fig. 15.1(b) we see how an irrational length a, taken as a limit of fractions $\frac{m}{n}$, leads to the same multiple A of the unit square. The theory of areas in elementary geometry along these lines is developed in detail by Moise [1974], who has the same difficulties with irrational segments as Euclid. Hartshorne, in his section 23, esp. 23.3, in effect uses VI.2 in order to prove VI.1.

15.3   The Basic Theorems of Similarity Geometry

The fundamental theorem VI.1 is immediately used to prove the theorem on proportional segments, Prop. VI.2, the universal tool in similarity geometry. (In the high-school text by Jacobs, VI.2 is called the "side splitter theorem." Jacobs reproduces Euclid's proof.)

Prop. VI.2.
If a straight line is drawn parallel to one of the sides of a triangle, it will cut the sides of the triangle proportionally; and, if the sides of the triangle are cut proportionally, the line joining the points of section will be parallel to the remaining side of the triangle.

For the proof Euclid uses the two auxiliary triangles $\triangle DEB$ and $\triangle DEC$ from Fig. 15.2(a).

Proof   (part 1)
Let $DE \parallel BC$. Then the two triangles with the common base DE have equal areas.

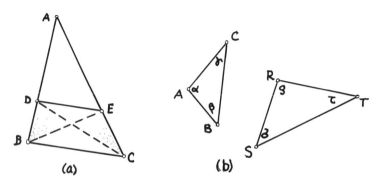

FIGURE 15.2

Now, prop VI.1 implies that

$$\triangle BDE : \triangle ADE \;=\; BD : AD,$$
$$\triangle CDE : \triangle ADE \;=\; CE : AE,$$

and because of $\triangle BDE = \triangle CDE$ we have

$$BD : DA = CE : EA.$$

Proof (part 2)

Euclid uses the proportion of lines and VI.1 to establish the equality of the two auxiliary triangles with the common base DE and then Prop. I.39 to derive $DE \parallel BC$.

The proposition about the angular bisector, VI.3, is a nice application of VI.2. The theory proceeds with Props. 4/5, which are converse to each other. We abbreviate the statements by using the notation from Fig. 15.2(b).

Propositions VI.4 and 5 say that

$$\left.\begin{array}{c} \alpha = \rho \\ \text{and} \\ \beta = \sigma \\ \text{and} \\ \gamma = \tau \end{array}\right\} \text{ if and only if } \left\{\begin{array}{c} b : c = s : t \\ \text{and} \\ c : a = t : r \\ \text{and} \\ a : b = r : s. \end{array}\right.$$

The essential point of this equivalence is that shape (as determined by the angles) can be expressed by means of proportions. At a time when incommensurable segments were unknown, the ratios of the segments could be expressed by numbers, and consequently, shapes of triangles—and more generally of figures composed from triangles—could be described by numbers. One might suspect that this theorem together with insights about harmonic intervals in music theory gave strong support to the Pythagorean doctrine that "everything is number." Clearly, nothing of this kind can be found in the *Elements*, which is a work with no metaphysical content.

Our formulation of VI. 4/5 contains deliberate redundancies in order to get a homogeneous and symmetric statement. The last lines could have been omitted, on the left side because of I.32 and on the right side because of the *ex aequali* property of proportions.

In analogy to the congruence theorems, Props. VI. 4/5 are called the *AAA* similarity theorem. (Sometimes *AA* theorem). In the same way Props. VI 6/7 are the *SAS*–similarity theorem. Prop. VI.8 treats an important special case.

Prop. VI.8.

If in a right-angled triangle a perpendicular is drawn from the right angle to the base, the triangles adjoining the perpendicular are similar both to the whole and to one another.

A direct sequel to Prop. VI.8 is Prop. VI.13 about finding a mean proportional to two segments.

Prop. VI.13.

To two given straight lines to find a mean proportional.

Proof: Place the given segments AB, BC in a straight line and draw the semicircle with diameter AC (Fig. 15.3). Let BD be perpendicular to AC and D on the semicircle. Because the angle in a semicircle is right, we have the similar triangles of VI.8 and hence

$$AB : BD = BD : BC.$$

Taking into account the possibility of cross multiplying a proportion established in Prop. VI.16/17, this proposition amounts to nothing else but Prop. II.14, the squaring of a rectangle. Just as in our discussion of Props. III.36 and IV.10 we have a simpler proof using proportions and a more complicated one avoiding proportions. We see again how the proofs of Books I–IV have apparently been intentionally rewritten older proofs with proportions.

From a passage in Aristotle it is obvious that he knew both proofs for the squaring of a figure. He discusses definitions and takes as an example:

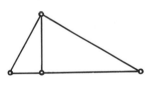 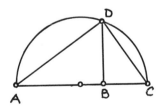

FIGURE 15.3

For instance, what is "squaring"? The construction of a square equal (in area) to a given rectangle. Such a definition is a statement of the conclusion, whereas, if you say that squaring is the finding of a mean proportional, you state the cause of the thing defined. (*De Anima* 413 a)

From the many mathematical examples in Aristotle's writings it is evident that he was well aware of and competent in the mathematics of his time, but he was not himself an active mathematician. If the passage above reflects discussions of the mathematicians among his friends, then we may speculate that this took place during his time in Plato's Academy (366–348). In any case, Aristotle must have learned about the two different proofs well before his death, in 322. Thus it is reasonable to assume that the version of Books I–IV without proportions was written in the interval 360–330, with possible later alterations in Book I concerning parallels and parallelograms. Proclus (Morrow p. 60) explicitly informs us about an author of "Elements" who *avoided proportion*. Euclid would then have edited the books, added his own proof of the theorem of Pythagoras, and incorporated them in his *Elements*. The key theorem for the transformation of statements about proportions into ones using areas is VI.16. In Books I–IV this remains veiled, but in hindsight the principle becomes transparent. The "new" proofs in I–IV look all the more ingenious after we have seen the simpler similarity proofs.

The rest of the propositions of Part B/Book VI show how to cut a segment similarly to a given cut segment and how to find third and fourth proportionals to given segments.

15.4   Book VI, Part C: Proportions and Areas (Products)

The main theorem of part C is Prop. VI.16. Euclid, like a modern author, puts the more technical parts of the proof in a lemma like Prop. VI.14 and adds some supplements in VI.15 and 17. In the figure an old friend from I.43 reappears. The whole part C does not depend on parts A and B of Book VI. For the proof Euclid returns to the roots VI.1. We state Props. 14 and 16 and sketch the proof.

Prop. VI.14.

In equal and equiangular parallelograms the sides about the equal angles are reciprocally proportional; and equiangular parallelograms in which the sides about the equal angles are reciprocally proportional are equal.

Prop. VI.16.

If four straight lines are proportional, the rectangle contained by the extremes is equal to the rectangle contained by the means; and, if the rectangle contained by the extremes is equal to the rectangle contained by the means, the four straight lines will be proportional.

We will replace the equiangular parallelograms by rectangles without any essential loss.

Euclid labels and places the parallelograms as in Fig. 15.4, such that the points D, B, E and F, B, G, respectively, are collinear. In our figure the "invisible" rectangle is completed.

Proof, part 1. Let the rectangles be equal. We denote the rectangles as does Euclid:

$$\square AB = \square BC \qquad \Rightarrow \qquad \square AB : \square FE = \square BC : \square FE.$$

Now we use VI.1:

$$\square AB : \square FE = \mathrm{line}DB : \mathrm{line}BE,$$
$$\square BC : \square FE = GB : BF.$$

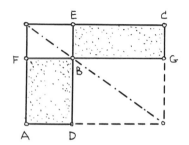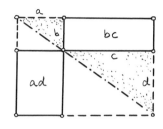

FIGURE 15.4

Hence

$$DB : BE = GB : BF,$$

the sides are reciprocally proportional.

Part 2. Reverse all steps of part 1.

Observe that the proof essentially uses the gnomon consisting of the rectangles $\square AB$, $\square FE$, and $\square BC$.

Prop. VI.16 is now an obvious reformulation of Prop. 14.

Because the equality of rectangles for Euclid means "of equal content" or "of equal area," we may use the modern notation a, b, c, d for (measures of) lines and ad, bc for (measures of) areas and restate VI.16 in the familiar form

$$a : b = c : d \qquad \Leftrightarrow \qquad ad = bc.$$

There are two more theorems about rectangles and similarity in part E of Book VI. Euclid postpones them to part E because he develops the concept of similar figures in part D. For the moment we will take similarity for granted and quote Props. VI.24, 26 in the context of the geometry of rectangles. In part E Props. 24, 26 play a subordinate role. The symbol $\approx$ denotes the similarity of figures (see Fig. 15.5).

Prop. VI.24:
$T \in AC \Rightarrow \square AT \approx \square TC \approx \square AC.$

Prop. VI.26:
$\square AT \approx \square AC \Rightarrow T \in AC.$

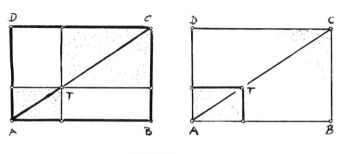

FIGURE 15.5

By restricting Euclid's proofs to rectangles instead of parallelograms, we have ignored another consequence of VI. 14/16. The two theorems together establish the fact that if two equiangular parallelograms are equal, they will remain equal if the angles are changed (in particular, to right angles) while the sides remain the same.

It is tempting to read into the group of theorems I.43, VI.14/16 and VI. 24/26 the remains of what Wilbur Knorr has called "the Doric geometry of rectangles" in contrast to "the Ionic geometry of triangles." It could be replenished by I.41–44 in the version for rectangles and some propositions from Books II and XIII. We have, however, no other clues for this conjecture than the correspondence with respect to the contents of the propositions.

15.5 Book VI, Part D: Similar Rectilinear Figures

Def. VI. 1.
Similar rectilinear figures are such as have their angles equal one by one and the sides around the equal angles proportional.

The first thing that Euclid does in this section is to show how similar figures can be constructed (Prop. VI.18). Having secured the existence of similar figures, he proceeds to the most important theorem, which connects lines and areas with respect to similarity:

Prop. VI.19.
Similar triangles are to one another in the duplicate ratio of the corresponding sides.

We recapitulate from Book V what a duplicate ratio is:

Def. V. 9.
If we have $a : b = b : c$, then $a : c$ is said to be the duplicate ratio of $a : b$.

We rewrite this in modern terms: If $a : b = b : c = k$, then $a : c = (a : b) \cdot (a : b) = k^2$.

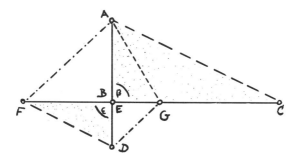

FIGURE 15.6

Read this way, Prop. VI.19 tells that if the sides of two similar tri-angles are connected by a similarity factor k, then their areas are related by the factor k^2. This is applied in VI.25 the other way round: A relation $a : c = k^2$ is postulated for areas, and the factor k has to be found as a mean proportional to a and c.

In the proof of VI.19 Euclid artfully manipulates proportions for lines until he can apply the fundamental theorem VI.1 in order to catch the areas.

We repeat Euclid's arguments but draw the corresponding figure (Fig. 15.6) in such a way as to make the procedure a little more transparent (at the expense of giving a point two labels, B and E).

We start with $\beta = \varepsilon$ and

$$AB : BC = DE : EF.$$

Alternation gives us

$$AB : DE = BC : EF.$$

Now construct the point G on BC such that

$$AB : DE = EF : BG.$$

From this we get two bonuses:

(1) (The area of) $\triangle ABG$ is equal to the triangle $\triangle DEF$ by VI.15 (or cross multiplication if the equal angles β, ε are right).

(2) $BC : EF = EF : BG$, so that $BC : BG$ is the duplicate ratio of $BC : EF$.

Now Prop. VI.1 secures the final result:

$$\triangle ABC : \triangle DEF = \triangle ABC : \triangle ABG = BC : BG,$$

as desired.

The next proposition, VI.20, extends the result of VI.19 to similar polygons by means of triangulations. Proposition VI.21 shows that the relation of similarity is transitive.

15.6   Book VI, Part E: The Application of Areas

Translated into modern mathematics, the Greek procedure of application of areas amounts to the solution of quadratic equations (or problems), where the coefficients and unknowns are represented by line segments. We have already discussed the most important special cases of this method in the chapters about Pythagoras and in the comments on Props. II 5/6. Because we know the essentials from these cases, we will not discuss the generalization from squares to similar parallelograms in VI. 28/29. Very detailed analyses can be found in the literature, cf. Notes. In the years around 1980 there was a heated debate about the right interpretation of the method of the application of areas. Some mathematicians have seen it as essentially the algebra of quadratic equations in geometric disguise, whereas historians insisted on a purely geometric reading of the texts. Certainly, a translation of involved geometric constructions into algebraic formulas is often helpful for the modern reader, as we have seen in various instances. But this is not to the point of the discussion. It is more about whether the Greeks were influenced by Babylonian algebraic methods and used geometry only as a disguise of algebra, or whether Greek mathematics was genuinely geometric and had nothing to do with algebra. In my opinion the two positions in the debate about geometrical algebra are a reflection of two quite different points of view, which might be called mathematical and philological. Mathematicians tend to stress isomorphisms; they like to see the same structure in different guises. By contrast, a philologist puts great value on expression and literary form. Mathematicians are accustomed to separating the content of a proposition from its form of expression, whereas philologists are likely to stress the particularity of different forms of expression. Since the history of Greek mathematics lives at the crossroads of these "two cultures," debates such as the one about geometric algebra are unavoidable. Indeed, they are essential to progress in our understanding of ancient science.

Proposition VI.25

Apollonius, in his *Conica* I.12/13, makes use of the application of areas with rectangular defect or excess instead of the square defect/excess that we have studied earlier. This forces some modifications on the methods used, most notably the introduction of Prop. VI.25. Remember that the theorem of Pythagoras can be seen in the following way: Construct an area of prescribed form, namely a square, of content equal to that of another figure, in this case consisting of two squares.

Prop. VI.25.

To construct one and the same figure similar to a given rectilinear figure and equal to another given rectilinear figure.

In the proof of this proposition Euclid takes the first given figure to be a triangle ABC. I take it to be a rectangle $ABCD$ without loss of generality. In the proof we will stick to rectangles instead of Euclid's parallelograms with prescribed angles. The construction proceeds as follows (see Fig. 15.7).

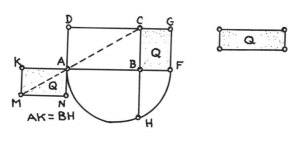

FIGURE 15.7

Let the given area be Q and the prescribed form be represented by the rectangle $\square$ *ABCD*. Apply the area Q to the line *BC* by means of Prop. I.44. Construct the mean proportional *BH* to the lines *AB* and *BF* using Prop. VI.13. (Algebraically, this amounts to extracting the root $BH = \sqrt{AB \cdot BF}$, which has to be used in VI 28/29.) Now the rectangle $\square$ *AKMN* on *AK* equal to *BH* will be "similar and similarly situated" to $\square$ *ABCD* and have area Q after VI 19/20.

Plutarch, a philosopher and popular writer in the philosophical tradition of Plato and Pythagoras who lived about 45–125 C.E. knew Prop. VI.25 as an important generalization of the theorem of Pythagoras. He writes:

> Among the most geometrical theorems, or rather problems, is the following: Given two figures, to apply a third equal to the one and similar to the other, on the strength of which discovery they say moreover that Pythagoras sacrificed. This is unquestionably more subtle and more scientific than the theorem which demonstrated that the square on the hypotenuse is equal to the squares on the sides about the right angle. (Plutarch *Symp.* VIII. 2,4 quoted after Euclid–Heath vol. I p. 343/44)

We will study other generalizations of the theorem of Pythagoras and say a few more words in the following chapter on generalizations.

Theorem VI.25 has a certain resemblance to Aristotle's doctrine (*Metaphys.* 1029a) of matter and form. A sculptor takes some matter—e.g., bronze—and gives it some form—e.g., a statue. As a certain amount of bronze can be given a variety of forms, in Prop. VI.25 an idealized matter, "area," can be cast into every polygonal form. Aristotle, to be sure, speaks only of the sculptor and some other examples, but not of Prop. VI.25. Moreover, we have to postulate an idealized substance "area," which might resemble a Platonic idea but not one of Aristotle's concepts.

A three-dimensional analogue of Prop. VI. 25 would, in its simplest application, solve the problem of the duplication of a cube. By Props. XI. 33 and XII. 8, 12, 18 (certain) similar solids are to one another in the triplicate ratio of their corresponding sides. Hence one would need a construction of a cube root (or, in Euclid's terminology, of two mean proportionals) for the solution of the solid problem corresponding to Prop. VI. 25. In the solid case, water could play the role of what we have called "area" in the plane case: To construct one and the same vessel similar to a given vessel and holding (exactly) a given amount of water. Curiously, the "ideal" plane problem is solvable by Euclidean means, but the "real" solid problem remains intractable.

There is a passage by Proclus that seems to support the interpretation of area as a sort of intelligible matter. Proclus discusses Euclid's Def. I.1 of a point and near the end mentions a certain doctrine of the Pythagoreans concerning the unit (one) of numbers and the point in geometry:

> Since the Pythagoreans, however, define the point as a unit that has position, we ought to inquire what they mean by saying this. That numbers are purer and more immaterial than magnitudes ... are clear to everyone. But when they speak of the unit as not having position, I think they are indicating that unity and number—that is, abstract number—have their existence in thought. . . . By contrast the point is projected in imagination and comes to be, as it were, _in a place and embodied in intelligible matter_. Hence the unit is without position, since it is immaterial and outside all extension and place; but the point has position because it occurs in the bosom of imagination and is therefore enmattered. (Proclus–Morrow p. 78, my italics)

Again Proclus does not speak of area, and we have to leave the finer points of this discussion to the philosophers and classical philologists.

16

The Origin of Mathematics 8

Be Wise, Generalize

One of the most essential features of a modern mathematical theory is its generality. Mathematicians try to present their results as generally as possible. Uniting seemingly disparate phenomena in a new theory earns great praise from fellow researchers. In this section we will study, in a few cases, the movement from a particular to a more general statement.

It is only through hindsight that we are able to analyze generalizations. There is no way of telling in advance how or in which direction to extend a particular result. This is left to the ingenuity of the particular mathematician.

There are various kinds of generalizations. Some work by abstraction of several cases. Others proceed by weakening the hypotheses of a theorem, but essentially preserving the arguments of the special case and arriving at a result similar to the previous one. Yet others start from a given statement and extend it in various directions. Then there may be two proposititions that come under a common roof in such a way that the original statements look like extreme cases. We will find examples of these types in Euclid's work and supply a few more from other authors.

16.1   Extending a Result

The sum of the angles in a triangle is two right angles. Euclid does not care to extend this result to polygons, but Proclus in his commentary on I. 32 does: A (convex) polygon may be divided into triangles, and hence the sum of the interior angles of a quadrangle is 2(2R), of a pentagon 3(2R), and so on. This same method of triangulation works in the case of VI. 19/20 concerning the relation of the areas of similar triangles/polygons. If we have a similarity factor k for lines, then the corresponding factor for the areas is k^2. The result in the general case is exactly the same as in the particular case. The proof is laborious, but essentially straightforward, and needs no specific new idea.

16.2   Weakened Hypotheses

Throughout Euclid's geometry, the experienced reader recognizes statements about parallelograms "with a given angle" that have their roots in the corresponding statements about rectangles. From a modern point of view, these generalizations are rather simple. If we shear a rectangle to a parallelogram on the same base and under the same height (Fig. 16.1), areas and ratios are preserved. Hence any statement concerning areas and proportions remains true when generalized from rectangles to "parallelograms with a given angle." The relation of triangles and parallelograms is more straightforward in certain cases (e.g., I. 41) than that of triangles and rectangles, but for most purposes, including the application of areas, the generalization to parallelograms seems to be pointless. Here, in particular, Euclid appears to be following the well-known postulate of the modern mathematician: "Be wise, generalize."

Another striking example of the weakened hypothesis, even if not presented by Euclid in this way, are the theorems about an-

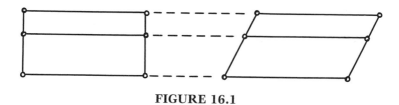

FIGURE 16.1

gles in circles. Thales has shown that the angle in a semicircle is right (Prop. III. 31). What about angles in other segments? Angles in the same segment of a circle are equal to one another (III. 21). From III. 20 we know a little more about the size of the angle in a segment, but essentially we have the typical situation: The assumptions are weakened, and so are the consequences, but they are still very useful.

16.3   Abstraction

The most prominent example of a generalization by abstraction is Euclid's theory of proportions for general magnitudes. We have seen the details of this in our discussion of Book V. Magnitudes of quite different kinds are treated according to one general concept. This much resembles the modern algebraic theories of groups, rings, fields, and similar algebraic structures.

Regular polygons were been treated by Gauss in a similar way. Instead of studying them case by case as Euclid did, he subsumed all of them in the abstract equation $z^n - 1 = 0$ and found the solution this way.

16.4   The Case of Pythagoras's Theorem

There is a host of generalizations of Pythagoras's theorem. We begin by listing the ones of the following type: Look in the direction of replacing the squares by other figures. Remember that Hippocrates of Chios had already done this in his treatment of the lunes when he considered semicircles on the sides of the right triangle. Euclid is more cautious in the following proposition about rectilinear figures.

Prop. VI.31:
In right-angled triangles the figure on the side subtending the right angle is equal to the similar and similarly described figures on the sides containing the right angle.

In Euclid's text the corresponding figure shows similar rectangles described on the sides. Readers may convince themselves of the equality $CX = DY$ of the similar rectangles. There is another generalization of this due to Pappus. Take any triangle $\triangle ABC$ and describe two parallelograms P_A, P_B on the sides BC and CA. Intersect the other sides of the parallelograms in point X as shown in Fig. 16.2. Extend XC and make $DY = XC$; draw the corresponding parallelogram P_C. Then P_C will be of equal content with P_A plus P_B. This is trivially true. I have never seen any application of it. It is what one might want to call a shallow generalization. It is so general that it is no longer of any use.

We have already seen the very fruitful generalization of Pythagoras's theorem in a second direction: Keep the squares but change the right angle, or in other words, calculate the third side of a triangle from the other two sides and their enclosed angle. One could view the resulting law of cosines as the generalization of three known special cases, namely for the angles $\gamma = 0°$, $\gamma = 90°$, and $\gamma = 180°$, with $c = AB$, $b = AC$, and $a = BC$ (cf. Fig. 16.3).

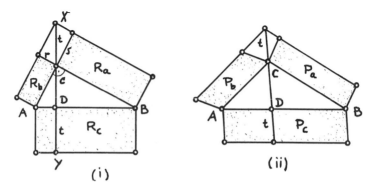

FIGURE 16.2

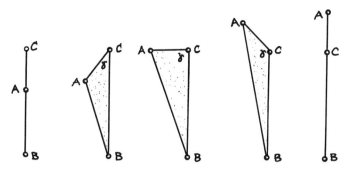

FIGURE 16.3

From the figures we get directly

$$
\begin{array}{llll}
\gamma = 0^\circ & : & c^2 = (a-b)^2 & = & a^2 + b^2 - 2ab, \\
\gamma = 90^\circ & : & c^2 & = & a^2 + b^2, \\
\gamma = 180^\circ & : & c^2 = (a+b)^2 & = & a^2 + b^2 + 2ab.
\end{array}
$$

For general γ we know that

$$c^2 = a^2 + b^2 - 2ab \cos \gamma,$$

and we regain the above formulas for $\gamma = 0^\circ, 90^\circ, 180^\circ$ as special cases of the general result.

16.5   The Generalization of Ptolemy

We can learn more about generalizations from Ptolemy (~ 150 C.E.), who generalized Pythagoras's theorem in his *Almagest* (Book I, Ch. 10), which was the basic astronomy textbook for many centuries. In order to understand Ptolemy, we have first to go back to a proof of Pythagoras's theorem that uses the methods of Book VI.

The right triangle $\triangle ABC$ in Fig. 16.4 is divided by its height h into two subtriangles as in Prop. VI. 8 with the result

$$
\begin{array}{lll}
a : c = & x : a & \Rightarrow \quad a^2 = cx, \\
b : c = & (c-x) : b & \Rightarrow \quad b^2 = c(c-x),
\end{array}
$$

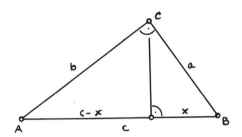

FIGURE 16.4

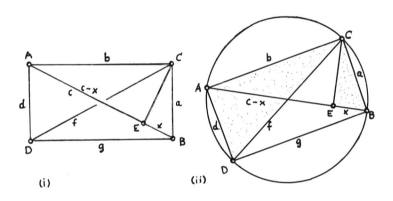

FIGURE 16.5

and hence

$$a^2 + b^2 = cx + c(c - x) = c^2.$$

For the intended generalization, this proof has to be transformed into a statement about rectangles. The right triangle reappears as half of a rectangle with diagonals c and f.

In the new context of Fig. 16.5(a), our proof from above reads:

$$\triangle EBC \approx \triangle ADC \;\Rightarrow\; x : a = d : f \qquad \Rightarrow\; xf = ad,$$
$$\triangle ECA \approx \triangle BCD \;\Rightarrow\; (c - x) : b = g : f \;\Rightarrow\; (c - x)f = bg.$$

Observing that $a = d$ and $b = g$ and $c = f$, we derive as above

$$
\begin{aligned}
xf &= & xc & = & ad & = & a^2, \\
(c - x)f &= & (c - x)c & = & bg & = & b^2,
\end{aligned}
$$

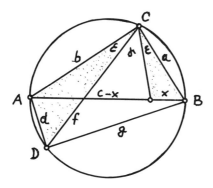

FIGURE 16.6

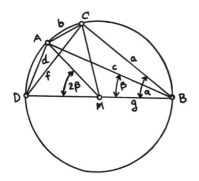

FIGURE 16.7

the theorem of Pythagoras for $\triangle ABC$. The ingenious idea of Ptolemy (or his predecessor) was to see that this proof would work for any quadrangle in a circle, if he only provided the similar triangles by making $\angle ECB = \angle ACD = \epsilon$.

Let $ADBC$ be a quadrangle in a circle (Fig.16.6). Place E on AB such that $\angle ECB = \angle ACD = \epsilon$. The angles $\angle ADC$ and $\angle ABC$ stand on the same chord AC and hence are equal. This makes $\triangle EBC \approx \triangle ADC$, and we get

$$x : a = d : f \Rightarrow xf = ad.$$

The triangles $\triangle EAC$ and $\triangle BDC$ have angles $\gamma + \epsilon$, and their respective angles $\angle BAC$, $\angle BDC$, subtend the same arc. They are therefore similar, providing as above

$$(c - x) : b = g : f \Rightarrow (c - x)f = bg.$$

Taking the two equations together, we derive the theorem of Ptolemy for quadrangles in circles:

$$cf = ad + bg,$$

which for a rectangle is nothing else but Pythagoras's theorem. This proof is an extraordinary piece of mathematical ingenuity and insight. Moreover, it has been of eminent importance for the further development of mathematics and its applications. This is because it provides us with the addition theorems for the trigonometric functions. Ptolemy used his theorem for the calculation of tables of chords, which we will replace by the more familiar corresponding values of the sine and cosine functions.

In Fig. 16.7 let BD be a diameter of the circle with radius r and a quadrangle $ADBC$ in this circle as shown. We have $DB = g = 2r$ and get

$$cf = ad + 2rb,$$
$$2rb = cf - ad.$$

In Fig. 16.7 we find the isosceles triangles $\triangle AMD$, $\triangle AMC$, $\triangle DMC$ with angles 2β, $2(\alpha - \beta)$, 2α respectively at M. This implies

$$\frac{f}{2} = r \sin \alpha, \qquad \frac{a}{2} = r \cos \alpha,$$
$$\frac{d}{2} = r \sin \beta, \qquad \frac{c}{2} = r \cos \beta.$$
$$\frac{b}{2} = r \sin(\alpha - \beta),$$

Putting these values for $a, b, c, \ldots$ into the equation above, we get

$$\sin(\alpha - \beta) = \sin \alpha \cos \beta - \cos \alpha \sin \beta,$$

the first of the addition formulas for the sine and cosine functions. The others are derived easily from this one. For many centuries, astronomers have used the tables calculated by Ptolemy for the sine and cosine functions (or rather the corresponding chords). In theoretical mathematics, the addition theorems provided the functional equation $e^{z+w} = e^z e^w$ for the complex exponential function $e^{x+iy} = e^x(\cos y + i \sin y)$. Later on, power series methods were used for easier access, but seen historically, the generalization of Pythagoras's theorem by Ptolemy led us far into modern mathematics.

Returning to the subject of generalizations on the whole, we notice how intricate this important generalization was. Up to the time of Euler, Ptolemy's theorem was more important than the law of cosines, the other fundamental generalization of Pythagoras's theorem. After Euler introduced the method of power series, Ptolemy receded into history. With the advent of vector spaces, the law of cosines and its modern equivalent, the scalar product for vectors, came into the foreground. Who can predict the fortunes of contemporary mathematical theories?

16.6 Marcel Proust on Abstraction

In "Time Regained," the last book of his monumental work *Remembrance of Things Past*, Marcel Proust develops his philosophy of art. In a most surprising way his artistic impulses as a writer are parallel to those of a mathematician who works on abstractions or generalizations. Proust writes:

> There was in me a personage ... coming to life only in the presence of some general essence common to a number of things, these essences being its nourishment and its joy. ...
>
> Just as a geometer, stripping things of their sensible qualities, sees only the linear substratum beneath them, so the stories that people told escaped me, for what interested me was not what they were trying to say but the manner in which they said it and the way in which this manner revealed their character or their foibles: or rather I was interested in what had always, because it gave me specific pleasure, been the goal of my investigations: the point that was common to one being and another. (Proust 1981 p. 737/38)

17

Euclid Book VII

Basic Arithmetic

....

17.1   The Historical Background

In *Prometheus Bound*, Aeschylus, the first of the three great classical writers, tells us how Prometheus taught humans not just the use of fire but also so many other worthwhile things. In particular, Prometheus relates, "And numbers, too, the chiefest of sciences, I invented for them" (lines 459/460). These lines, written about 465 B.C.E., bear witness to the high esteem the early Greeks had for arithmetic.

There is, however, a complementary way of writing about the origins of cultural achievements. Aeschylus represents the more mythological side of Greek thinking, whereas Herodotus, who was writing a few years later (about 440) stands for the more practical, we might even say enlightened, way of writing historical texts. Eudemus tells us that the Greeks learned geometry from the Egyptians and arithmetic by the way of commerce from the Phoenicians. Herodotus relates us this story about geometry (*Historia* II.109) some 150 years before Eudemus. Herodotus does tell us a little more about

161

arithmetic. He observes that Greeks and Egyptians calculate by using pebbles (Greek *psephoi*, Latin *calculi*) on a calculating board or abacus (*Historia* II.36). This agrees very well with the practice ascribed to the Pythagoreans of doing arithmetical investigations with the help of pebbles. In the context of another story, Herodotus calculates that a 70-year-old person has lived a total of about 25,200 days (and not a single one having been like any other of them).

Coming back to practical questions, we note that currency exchange rates must have played a considerable role in Greek economics. Money was introduced around 600 B.C.E., and its use spread gradually among the Greek city-states. Weight standards for silver coins varied among groups of city-states, creating the need among businessmen for a mechanism for exchanging the different monies.

There is an allusion to this sort of banking business in the New Testament, Matthew 21: 12: "And Jesus went into the temple of God, and cast out all of them that sold and bought in the temple, and overthrew the tables of the money changers, and the seats of them who sold doves." People from all over the world gathered for holy days in Jerusalem. We can easily imagine similar commercial activities and the presence of money changers at the great Greek festivals at Delphi or Olympia.

Exchange rates are not really ratios, but proportions. Let us illustrate this by looking at recent quotations for the present-day currencies US dollars, Swiss francs, and German marks (DM). We read in the paper

$$DM \; : \; US\$ \; = \; 173 \; : \; 100,$$
$$US\$ \; : \; SF \; = \; 100 \; : \; 145.$$

From this we can easily derive, using the Greek method of *ex aequali*,

$$DM \; : \; SF \; = \; 173 \; : \; 145.$$

In the time 580–480 B.C.E. the coinage of Aegina was the leading currency (like the US $) around the Aegean sea, and transactions like the one above are easily imaginable. Aristotle has a similar example in his *Nicomachean Ethics* (1133–20–27), where the relative values of different commodities are measured using money.

We do not know how much of Euclid's arithmetic is indebted to commerce. However, we will see that in some of its aspects, it has strong relations to geometry. What we find in Euclid's arithmetical books is mathematics for its own sake like the geometrical parts of the *Elements*.

Very faint traces of the Pythagorean pebble arithmetic can be seen in Euclid's books. We will return to this subject in the section on Iamblichus and Diophantus. It is appropriate to look at Books VII, VIII, and IX as a self-contained part of the *Elements*. The main topics treated in this part are foundations (proportions, greatest common divisor), sequences of numbers in geometric proportion, the geometry of numbers (plane and solid numbers, squares and cube), and, at the end, the elementary theory of odd and even numbers.

17.2 The Overall Composition of Book VII

Definitions	
1–4	A: The Euclidean algorithm
5–7	B: Basic statements about proportions of numbers using Def. 20
11–16	C: Transformation of 5–7 into statements using the word "proportion"
17–19	D: Proportions and products
20–33	E: Theory of the greatest common divisor, prime divisors
34–39	F: Theory of the least common multiple

17.3 Definitions

The definitions at the beginning of Book VII are intended to serve for all of VII–IX. The first five definitions are descriptions of the basic concepts of unit, of number, and of "measuring" numbers; these definitions are analogous to the descriptions of point, line, etc. in Book

I. For Euclid, a number is always a natural number (integer) greater than one. Definitions 6 to 10 concern even and odd numbers, which clearly belong to the theory of the even and the odd in the second half of Book IX. Primes, composite numbers, and multiplication are defined in 11–15. Geometric shapes of numbers (e.g., "squares" and "cubes") are the subject of 16–19 and 21. In Defs. 16–19 multiplication is understood to be a symmetric operation, whereas in Def. 15 multiplication is defined in an asymmetric way. The group Definitions 16–19 seem to be older than Def. 15. Euclid tries to reconcile Defs. 15 and 17 in Prop. IX. 7. Proportionality of numbers is defined in 20, and finally, 22 says that a perfect number is the sum of its factors other than itself; this definition is invoked only at the end of Book IX.

In contrast to Book I, there are no postulates in a material sense in Book VII. Addition of numbers is taken for granted with all its properties. Although multiplication is defined in Def. 15, "measuring" (i.e., dividing) is assumed to be known in Def. 3 and obviously regarded as a more fundamental concept. We quote a selection of definitions that are either of fundamental interest or will be the most important ones for us in the subsequent text.

Definitions

Def. 1. *A unit is that by virtue of which each of the things that exist is called one.*

Def. 2. *A number is a multitude composed of units.*

Def. 3. *A number is a part of a number, the less of the greater, when it measures the greater;*

Def. 4. *but parts when it does not measure it.* (The meaning of Def. 4 will be explained in Prop. VII.4)

Def. 5. *The greater number is a multiple of the less when it is measured by the less.*

Def. 11. *A prime number is that which is measured by a unit alone.*

Def. 12. *Numbers prime to one another are those which are measured by a unit alone as a common measure.*

Def. 13. *A composite number is that which is measured by some number.*

Def. 15. *A number is said to multiply a number when that which is multiplied is added to itself as many times as there are units in the other, and thus some number is produced.*

Def. 16. *And, when two numbers having multiplied one another make some number, the number so produced is called plane, and its sides are the numbers which have multiplied one another.*

Def. 17. *And, when three numbers having multiplied one another make some number, the number so produced is solid, and its sides are the numbers which have multiplied one another.*

Def. 18. *A square number is equal multiplied by equal, or a number which is contained by two equal numbers.*

Def. 20. *Numbers are proportional when the first is the same multiple, or the same part, or the same parts, of the second that the third is of the fourth.*

Def. 21. *Similar plane and solid numbers are those which have their sides proportional.*

Def. 22. *A perfect number is that which is equal to its own parts.*

A short look at some of the definitions reveals certain inhomogeneities in the text. There are several historical layers, but we don't have the means at hand to extract a clear picture of the contributions of the various authors who preceded Euclid.

Definition 3 presupposes that we know what it means for one number to measure (divide) a greater one, and in Def. 5, a multiple is defined in these terms. In Def. 15, however, multiplication is introduced as a nonsymmetric operation reduced to addition, like the scalar–vector multiplication of magnitudes by numbers in Book V. In contrast to this, Def. 16 speaks of plane numbers produced by two numbers multiplying one another in an obviously commutative operation. We will have a closer look at these problems below.

17.4 Book VII, Part A: The Euclidean Algorithm

Euclid's famous algorithm is the basis of his number theory. He uses it to determine of the greatest common divisor (common measure in his terminology) just as it is used today. In this respect, nothing has changed in the intervening 2300 years. The Euclidean algorithm consists of repeated execution of division with remainder, which is still among the first things learned in a modern course on number theory. Written down in a formal way, division with remainder means:

(DR) Let a, b be natural numbers and $b > a$. Then there exist uniquely determined integers q and r such that

$$b = qa + r \quad \text{and} \quad 0 \leq r < a.$$

Euclid takes (DR) for granted, as well as the following statement: If d measures (divides) a and b, then d will measure $b \pm qa$.

Prop. VII.1. (The Euclidean algorithm)
If two unequal numbers are set out and the lesser is always subtracted in turn from the greater, then, if the remainder never measures the number before it until a unit is left, the original numbers will be prime to each other.

In his typical manner, Euclid abbreviates the successive subtraction in his proof to a few steps. Let $b > a$ be the two given numbers and let

$$b = qa + r, \quad r < a,$$
$$a = sr + t, \quad t < r,$$
$$r = ut + 1.$$

If a and b had a common measure $e > 1$, then e would measure r, hence also a and r and consequently t and r, so that it would measure 1, which is impossible. Hence a and b are prime to each other.

Prop. 2.
Given two numbers not prime to one another, to find their greatest common measure.

The proof is done in the same way as that of Prop. 1 with the obvious modifications. In the following Prop. 3, the greatest common measure is determined for three numbers.

The Greek word for the Euclidean algorithm seems to have been *anthyphairesis* (or *antaneiresis*), which means something like mutual subtraction. We will use it for segments in our section on incommensurability.

17.5   Book VII, Parts B and C: Proportion for Numbers

In order to get a better understanding of Def. 20, we will discuss it together with

Prop. VII.4:
Any number is either a part or parts of any number, the less of the greater.

This statement looks curious, because in Euclid's language "parts" means a number of equal parts, and if we take the unit as a part of the greater number a, then the smaller number b will certainly be "parts" of the greater. What is really meant is revealed by the proof of Prop. 4. Three cases are considered. (We use modern notation.) Let $a > b$ be the numbers in question.

(i) If a and b are relatively prime, then b is divided into b units, each of which is a part of a; hence b is "parts" of a.

(ii) If b measures a, then b is a part of a.

(iii) If b does not measure a and a and b are not relatively prime, then let g be the greatest common measure (divisor, gcd) of a and b and let $a = kg$ and $b = ng$. Then b is a number of parts of a, namely n parts.

Now let us look back at the definition of proportion for numbers.

Def. VII.20:
Numbers are proportional, when the first is the same multiple, or the same part, or the same parts, of the second that the third is of the fourth.

We translate into modern formulas. Let the four numbers be a, b, c, d.

We have $a : b = c : d$ if

$$a = nb \quad \text{and} \quad c = nd,$$

or

$$a = \frac{1}{k}b \quad \text{and} \quad c = \frac{1}{k}d,$$

or

$$a = k\frac{1}{n}b \quad \text{and} \quad c = k\frac{1}{n}d.$$

How can we find out whether $a : b = c : d$ is true? Seen in the light of Prop. VII.4, we have to determine the greatest common divisors $g = gcd(a, b)$ and $f = gcd(c, d)$ and numbers k, n and l, m such that

$$a = kg = k\frac{1}{n}b$$

and

$$c = lf = l\frac{1}{m}d.$$

We will then have $a : b = c : d$ if (and only if)

$$k = l \quad \text{and} \quad n = m.$$

Again rewritten and modernized into fractions:

$$\frac{a}{b} = \frac{kg}{ng} = \frac{k}{n} \quad \text{in least terms,}$$

$$\frac{c}{d} = \frac{lf}{mf} = \frac{l}{m} \quad \text{in least terms.}$$

In order to find out whether $a : b = c : d$, we have to know that for each fraction $\frac{a}{b}$ there is a *uniquely determined reduced fraction* $\frac{k}{n} = \frac{a}{b}$. This fact is hidden in Euclid's definition of the proportionality for numbers. If the uniqueness of $\frac{k}{n}$ were not assumed, one could not find out whether $a : b = c : d$.

Euclid develops a theory of what he calls "the least numbers of those which have the same ratio with them" in Section E (Props. 20–33) of Book VII. We will find the same assumption of existence and uniqueness as above in Section E. Only in Prop. VII.33 does he show how to find the "least in the same ratio" by using the *gcd* as we did above. Given the position of this proposition at the end of the section and its generality (for as many numbers as we please . . . to find the least ones in the same ratio), it appears to me to be a later interpolation or addition to the main text.

Propositions About Proportions in Terms of Def. 20.

Prop. VII.5.

If a number is part of a number and another is the same part of another, both together will be the same part of both together as the one is of the one.

Again, for the convenience of the modern reader, we will denote the numbers etc. in the way familiar today.

$$\text{If} \quad a = \frac{1}{n}b \quad \text{and} \quad c = \frac{1}{n}d,$$
$$\text{then} \quad a + c = \frac{1}{n}(b + d).$$

In the next proposition, this is extended to "parts" $\frac{m}{n}$ instead of $\frac{1}{n}$. Because of the logical chain starting from VII.5 and leading to the commutativity of multiplication in VII.16, we will have to have a closer look at the proof of VII.5, again in modern notation.

The proof of VII.5:

Let $a = \frac{1}{n}b$ and $c = \frac{1}{n}d$.
  Then

$$b \;=\; a + a + \cdots + a \qquad (n\text{times}),$$
$$d \;=\; c + c + \cdots + c \qquad (n\text{times}).$$

Hence, adding "columns,"

$$b + d = (a + c) + (a + c) + \cdots + (a + c) \qquad (n\text{times}),$$

that is,

$$a + c = \frac{1}{n}(b + d).$$

Clearly, the commutative law for the addition of numbers is used tacitly in this proof. We will see better what is going on if we let $a = c$. Then the above calculations result in

$$
\begin{aligned}
b + d &= na + na &= 2(na), \\
&= n(a + a) &= n(2a).
\end{aligned}
$$

Later on, in VII.9/19, Euclid uses VII.5/6 for any number k of summands. This amounts to

$$k(na) = n(ka),$$

and the commutative law VII.16 of multiplication becomes clearly visible. The logical chain leading to VII.16 is

$$\text{VII.5/6} \Rightarrow \text{VII.12} \Rightarrow \text{VII.15} \Rightarrow \text{VII.16}.$$

Another sequence leads from 5/6 to alternation of proportions in VII.13:

$$\text{VII.5/6} \Rightarrow \text{VII.9/10} \Rightarrow \text{VII.13}.$$

From the introduction to Book V we know how alternation is related to commutativity.

What we see behind the scenes is a rectangular array of a's with n columns and k rows:

$$
\begin{array}{ccccccc}
a & a & . & . & . & . & a \\
a & . & . & . & . & . & a \\
a & . & . & . & . & . & a \\
a & a & . & . & . & . & a
\end{array}
$$

Depending on how one counts, one gets $k(na)$ or $n(ka)$. If a is taken to be the unit, this will be the commutative law. In VII.11–14 the results of 5–10 are translated from the language of parts into the language of proportions. The latter ones are very similar in their wording to the corresponding propositions in Book V. For instance, VII. 12 is almost word for word the same as V, 12, just "number" and "magnitude" are interchanged. VII.13 is alternation (like V, 16), and VII.14 *ex aequali* corresponds to V. 22.

In spite of these similarities and more between propositions and proofs for numbers and segments later on, the two theories of proportion for numbers and magnitudes (or segments) seem to have different roots. All our sources from antiquity stress the difference between discrete numbers and divisible magnitudes like segments. Plato, for instance, speaks in detail about the "limited" (number) and the "unlimited" (magnitude) in the *Philebus* (24 sq., esp. 25ab). There may have been an attempt for a unified theory of proportion using the Euclidean algorithm for segments as well as for numbers (cf. Fowler [1987]). We will speak about this in connection with Book X and the study of incommensurables. Otherwise, there is no evidence left for any definition of proportionality other than the ones we have in Euclid's Books V and VII. So we just cannot tell what has been replaced by Def. VII 20, even if this and the corresponding Props. VII.5–10 do indeed look like a new underpinning for a theory that did exist considerably earlier.

17.6 Book VII, Part D: Proportions and Products

This is a short but important section consisting of three propositions only. It is closely related to some equally important theorems from Book VI. Before going into the details, a few remarks about Euclid's representation of numbers by segments are in order.

In the diagrams that go with most of the propositions in the arithmetical books VII–IX, numbers are represented by segments. These segments obviously play the role of modern variables. Sometimes a number/segment is denoted by a single letter, like A, and even the unit is represented (e.g., in Prop. VII.16) by a segment and letter E. When the number has to be divided later on in the proof, the two endpoints of the segment are labeled, and subsequently other letters then are introduced to denote the parts (e.g. VII.15 where the number EF is represented as $EF = EK + KL + LF$). Products of numbers are depicted by segments like the factors (VII.16), even if an arrangement of dots as we had it in the last section and the analogy to the geometric books would suggest a rectangular representation. Later on, e.g., in VIII.18, Euclid speaks of rectangular numbers but

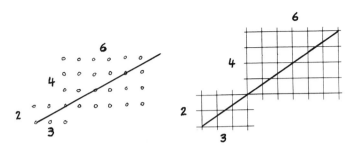

FIGURE 17.1

still uses segments for the pictorial representation. Let us look at two modes of graphical representations of numbers, even if neither of them is present in Euclid's *Elements*. First let a number be represented by an array of dots as suggested by the Pythagorean tradition. (Compare the quotations from Nicomachus in the next chapter.) A rectangular array is helpful for the visualization of a simple identity like $nk = kn$, but it breaks down for a slightly more complicated issue such as the use of a common diagonal for the visualization of proportions, see Fig. 17.1 with the example $3 : 2$ and $6 : 4$.

If, on the other hand, one uses a unit segment for the number unit, then the geometrical arguments from Book VI carry over to numbers almost effortlessly. That such a representation was used in Greek mathematics can be seen from the following quotation from Plato, where he calls 3 and 5 oblong (i.e., rectangular in contrast to square) numbers.

Theaetetus speaks, after he has defined square numbers:

> Any intermediate number, such as three or five or any number that cannot be obtained by multiplying a number by itself, but has one factor either greater or less than the other, so that the sides containing the corresponding figure are always unequal, we likened to the oblong figure, and we called it an oblong number. Socrates: Excellent. (Plato, Theaetetus 147e9–148a5)

After Theaetetus, a prime number like 5 has factors/sides of the corresponding rectangular figure.

Square grids have been used by artists at least since the Middle Kingdom (1800 B.C.E.) in Egypt for the transfer of sketches to the wall

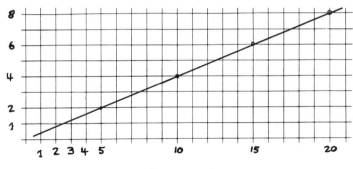

FIGURE 17.2

or for the design of sculptures. There is even a sketch of a rectangular grid in one of the manuscripts of the *Elements*, but not in relation to arithmetic.

Figure 17.2 shows how "the least of the numbers in the same relation with them" for short *minimi* (after Taisbak) could be seen in a square grid. Let me stress, however, that nothing of this kind is in the text of the *Elements*. Even if it is plausible that many of the proofs about numbers are geometrically motivated, no arithmetical theorem is ever proved by means of theorems from geometry (with some exceptions discussed in Chapter 20). On the surface, arithmetic is self sufficient.

Prop. VII.17:
If a number by multiplying two numbers makes certain numbers, the numbers so produced will have the same ratio as the numbers multiplied.

Proof (abbreviated)
The claim is

$$ab : ac = b : c.$$

Since a by multiplying b made ab, the number b measures ab according to the units in a. Hence we have

$$1 : a = b : ab \qquad \text{(by Def. VII.20)},$$

and similarly,

$$1 : a = c : ac.$$

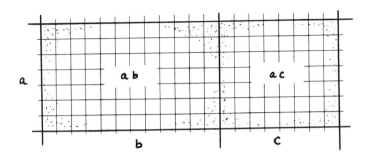

FIGURE 17.3

From this we get

$$b : ab = c : ac,$$

and by alternation, VII.13:

$$b : c = ab : ac.$$

If we look at the geometrical representation in Fig. 17.3, we see that VII.17 for numbers has essentially the same meaning as VI.1 for segments. In fact, VII.17 is used in the proportion–product Theorem VII.19 in exactly the same way as VI.1 in the corresponding theorem for segments/rectangles, VI.16, as we will see presently. The big difference is in the proofs of VI.1 and VII.17. For arbitrary (divisible, continuous) magnitudes the difficult theory of Book V is indispensible, while for the discrete case of numbers a much simpler proof is possible.

Observe again that the proof of VII.17 essentially consists in counting the small squares in Fig. 17.3 in two different ways. This is reflected in the use of alternation, which in many cases replaces commutativity in Euclid's arguments.

Proposition VII.18 is the version $b : c = ba : ca$ of VII.17 and is reduced to the latter by using the commutative law established in VI.16.

Prop. VII.19 (Proportion–product theorem, modern notation)
For any numbers a, b, c, d:

$$a : b = c : d \qquad \Leftrightarrow \qquad ad = bc.$$

This is the arithmetical version of the proportion–product theorem VI. 14/16 in geometry. (For the proof observe that Euclid, having to work without a convenient notation for products, uses $ad = E$, $bc = F$, and the auxiliary number $ac = g$.)

Proof, part (i): From VII.17/18 we get

$$a : b = ac : bc \qquad \text{and} \qquad c : d = ac : ad.$$

Now, from $a : b = c : d$ we infer

$$ac : bc = ac : ad,$$

and from this $bc = ad$ is immediate.

For part (ii), reverse all steps in part (i) of the proof.

There is a curious, but revealing, little line in the proofs of VI.16 and VII.20. In both cases Euclid says at the beginning of part (ii) of the proof,

"For, with the same construction ..."

and continues with equal rectangles in Book VI and with equal numbers ($ac = bc$) in Book VII. What should the construction be in Book VII? It is the construction of a gnomon like the one in Book VI (Fig. 17.4).

There is a fragment by the Pythagorean philosopher Philolaos of Croton that is discussed in detail by Burkert [1972, p. 273] and that is related to the above remarks. Philolaos himself lived around 450 B.C.E., but the fragment includes so many Platonic and Aristotelian allusions that most likely it is spurious. Even if it was written by

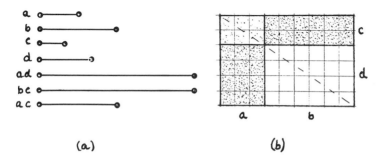

(a)   (b)

FIGURE 17.4   (a) Euclid's figure (b) Gnomon

some other author around, 300 B.C.E., it sheds some light on to the use of geometric methods in arithmetic: Number causes

> in the mind, in accordance with perception, everything to be knowable and conforming to each other in the way of a gnomon, composing and separating the ratios of things, separately the finite and infinite ones. (Philolaos fragment B 11 Diels–Kranz)

"The finite and infinite ones" has to be read as "the limited and the unlimited" or as "discrete and nondiscrete ones" as in other contexts of this kind. The fragment would then indicate the use of the gnomon in arithmetic as well as in geometry. Whatever the value of this fragment may be, the evidence for use of geometric ideas for arithmetical proofs will become overwhelmingly clear in the course of our discussion of Book VIII.

17.7 Book VII, Part E: The Greatest Common Divisor and Prime Divisors

The preparatory sections of Book VII are complete with Prop. VII.19. Basic number theory starts with Prop. VII.20. After quoting this proposition we will discuss the involved concept of "minimi."

Prop. VII.20.
The least numbers of those which have the same ratio with them measure those which have the same ratio the same number of times, the greater the greater and the less the less.

The fundamental importance of this theorem has been worked out by M. Taisbak [1971]. As mentioned earlier, we will follow Taisbak and call "the least of those which have the same ratio with them" the *minimi* in the collection of all pairs of numbers that have a certain given ratio. Apparently, Euclid takes the existence and uniquenes of the minimi in a class of pairs of numbers with the same ratio for granted, as we have already observed in our discussion of Def. VII.20.

The proof of Prop. 20 is as follows: Let r, s be minimi and $a : b = r : s$. We have to find a number n such that $a = nr$ and $b = ns$. By hypothesis, we have $r \leq a$ (and $s \leq b$), and by Prop. VII.4, r is either part or parts of a, that is, $r = \frac{1}{n}a$ or $r = \frac{k}{n}a$ with $k > 1$. The last possibility has to be ruled out. If $r = \frac{k}{n}a$ and $k > 1$, then $s = \frac{k}{n}b$ as well, and $r' = \frac{1}{n}a$ would be smaller than r, and together with $s' = \frac{1}{n}b$ would constitute a pair r', s' such that $r' : s' = a : b$ and $r' < r$ and $s' < s$, in contradiction to r, s being minimi.

Observe that as in geometry, Euclid does not speak explicitly about any properties of ordering. In this case he even has a sort of ordering of pairs of numbers, suggesting a picture like Fig. 17.2 or something similar for a class of pairs of numbers in the same ratio.

In the next propositions we obtain more information about minimi:

Prop. VII.21.
Numbers prime to one another are minimi.

Prop. VII.22.
Minimi are prime to one another.

There is one more basic proposition about minimi, which comes at the very end of part E of Book VII just before the theory of the least common multiple starts. It looks very much like a later addition to the original part E. In fact, only in Prop. VII.33 are we told how to find minimi, and this not only for pairs of numbers, but for as many numbers as we please. In this general aspect VII.33 also serves later developments in Books VIII/IX, but from its content it clearly belongs to one group of propositions with VII. 20, 21, 22.

Prop. VII.33.
Given as many numbers as we please, to find the least of those which have the same ratio with them.

As usual, Euclid takes three numbers for "as many as we please." We will be content with two numbers a, b. If $gcd(a, b) = 1$, then a, b are prime to one another and are minimi by VII.21. If $gcd(a, b) = d > 1$, then for some numbers u, v, we have $a = ud$ and $b = vd$. By VII.17/18 we may cancel d and have $a : b = u : v$. Now let r, s be minimi for a, b. Then Euclid shows that $r = u$ and $s = v$. Thus again the existence and uniqueness of minimi is assumed from the outset. We are merely told how to find them.

In the next group of Propositions, VII.23–28, Euclid explores how the concept of being relatively prime for two numbers a, b is related to forming products and sums. We again use modern notation and abbreviate "relatively prime" by $gcd(a, b) = 1$. In this shorthand the systematic development of 23–28 becomes most obvious.

VII, 23: $gcd(a, b) = 1$ and c divides $a \Rightarrow gcd(c, b) = 1$.

VII, 24: $gcd(a, b) = 1$ and $gcd(c, b) = 1 \Rightarrow gcd(ac, b) = 1$.

VII, 25: $gcd(a, b) = 1 \Rightarrow gcd(a^2, b) = 1$.

VII, 26: $gcd(a, b) = 1 \Rightarrow gcd(a^2, b^2) = 1$.

VII, 27: $gcd(a, b) = 1 \Rightarrow gcd(a^3, b^3) = 1 = gcd(a^n, b^n)$.

VII, 28: $gcd(a, b) = 1 \Leftrightarrow gcd(a + b, b) = 1$.

On this occasion Euclid again teaches us much more about mathematics than some simple facts. He shows us how in a systematic exposition one has to investigate how a new concept—in this case "relatively prime"—is related to preceding concepts and operations— in this case multiplication and addition. It is quite the same if one shows, e.g., in a calculus course, how the concept of continuity is related to the multiplication and addition of functions.

Prime Divisors

In the last part of Section E, Props. VII 29–32, two lemmas are proved whose combination is known today as the fundamental theorem of arithmetic: *Every number $n > 1$ has a unique factorization into prime numbers.* We will see how close Euclid's propositions come to this theorem, but in fact, it is *not* in the *Elements*.

Prop. VII.29.
Any prime number is prime to any number which it does not measure.

Let p be a prime number, and let it not measure a. If p and a had a common measure $c > 1$, then this would have to be smaller than p and measure p, which is impossible.

Prop. VII.30.

If two numbers by multiplying one another make some number, and some prime number measures the product, it will also measure one of the original numbers.

Euclid's proof is as follows: Let a, b be the numbers multiplied and let the prime p divide (measure) ab. It has to be shown that p divides a or p divides b. Because p divides $c = ab$, there is a number e such that $pe = c = ab$. From $pe = ab$ we get $p : a = b : e$ after the proportion–product theorem VII.19. If p divides a, nothing remains to be shown. If p does not divide a, then by the last proposition we have $gcd(p, a) = 1$, and by VII.21 the numbers p, a are minimi. Now, the essential step of the proof is the application of VII.20, to the effect that because p, a are minimi, there is a number n such that $np = b$ and $na = e$; hence p divides b.

In modern treatments of number theory, Prop. VII.30 is called the prime divisor property (PD). With the usual convention $x|y$ for x divides y it reads:
(PD) p prime and $p|ab \Rightarrow p|a$ or $p|b$.

(A short proof of (PD) based on the Euclidean algorithm is standard in courses on number theory.)

Prop. VII.31.

Any composite number is measured by some prime number.

Proof
Let the number a be composite. Since it is composite, some number b will measure (divide) it. If b is prime, we have found what we were looking for. If not, some other number c ($\neq 1$) will divide b and hence also a. If c is prime ... and so on.

Euclid concludes: "Thus, if the investigation is continued in this way, some prime number will be found which will measure the number before it, which will also measure a.

"For, if it is not found, an infinite series of numbers will measure the number a, each of which is less than the other: This is impossible in numbers."

In the hands of P. Fermat (1601–1665) this method of "infinite descent" became a famous principle of proof in number theory. It can be transformed into the method of mathematical induction, but there are no traces of a proof by induction in Euclid's work.

The next proposition is an easy extension of the last one:

Prop. VII.32.
Any number is either prime or measured by some prime number.

An immediate consequence of these propositions is the factorization of any number a into a product of prime numbers. When we write this down,

(F) $a = p_1 \cdots p_n$,

we realize in the same moment that Euclid lacked the formal means for such a statement. Anyway, he invented many clever things and could have found a substitute for this. In the beginnings of a modern course in number theory, one would use VII.32 for the existence and (PD) in a slightly extended variant in order to prove the uniqueness of (F). Let

$$p_1 \cdots p_n = a = q_1 \cdots q_r$$

be two representations of a as a product of primes. By (PD), the prime p_1 will divide one of the primes q_i, and hence $p_1 = q_i$. Cancel p_1 and q_i and proceed similarly with p_2. In the end, after some rearrangement of the q_j, one will get $n = r$ and $p_j = q_j$ for $j = 1, \ldots, n$. This completes the proof of the fundamental theorem of arithmetic based on VII 30. 32.

Why did Euclid (or any other Greek mathematician) not prove the fundamental theorem? This is in a way the same question as, Why didn't the Greeks invent the bicycle? Given their ability to build wonderfully efficient chariots for horse racing, the technology of building something like the first Draisine of K. Drais (1817) was certainly in their hands—but they just didn't do it, and so they just didn't think of stating and proving the fundamental theorem of arithmetic. The first mathematician who cared to prove the uniqueness part of the fundamental theorem was C. F. Gauss in his *Arithmetical Investigations* of 1801, article 16.

There is, however, one theorem in the *Elements* that comes rather close to the fundamental theorem. It reads:

Prop. IX, 14.

If a number is the least measured by prime numbers, it will be measured by no other prime number outside of those originally measuring it.

Before we look at the proof, let us observe two things: (i) Euclid gets around a product $a = p_1 \cdots p_n$ by saying that the p_i *measure* a certain number a. (ii) He speaks of the *least* number measured by certain prime numbers. This implies that every prime p_i occurs only once as a factor of a, or, in a modern phrase, that a is square–free. Even if he has found a way out of his dilemma in (i), he still will not say something like p^α measures a.

It is characteristic that the proof of Prop. IX, 14 uses the prime divisor property (PD), VII.30, in a way similar to our proof of the fundamental theorem above. It is as follows: Let a be the least number divided by the primes b, c, d. It has to be shown that no other prime e divides a. Assume that a prime e different from b, c, d divides a and $a = ef$, where $f < a$. By the prime divisor property (PD), each of b, c, d will divide e or f. Now, they cannot divide e by assumption; hence they have to divide f, which is less than a. Contradiction.

As the proof goes, it really depends on a being square free; otherwise one could not get the contradiction.

The two propositions preceding IX, 14 are, like IX. 14, concerned with what we would call special cases of the fundamental theorem. A statement using modern notation is as follows:

Prop. IX. 12.
p divides $a^n \Rightarrow p$ divides a.

Prop. IX. 13.
The only divisors of p^n are $p, p^2, \ldots, p^{n-1}$. (1 and p^n are not counted as divisors.)

In the proof of IX. 12, Prop. VII.20 is used in the same way as in the proof of (PD), but (PD) itself is not quoted. The proof of IX. 13 depends on IX. 12.

17.8   Book VII, Part F: The Least Common Multiple

In this part of Book VII, the least common multiple of two or more numbers is constructed, and its properties are investigated. No important new insights into Euclid's way of thinking would be gained by reporting the details.

18

CHAPTER

The Origin of Mathematics 9

Nicomachus and Diophantus

Aside from Euclid, there are two mathematicians from antiquity whose books about arithmetic have survived.

18.1 Nicomachus: *Introduction to Arithmetic*

Nicomachus of Gerasa lived around 100 C.E. He probably studied at Alexandria, at that time the center of mathematical studies and of neo–Pythagoreanism. It may be that he preserves some Pythagorean doctrines about numbers in his writings. His mathematical style is quite different from Euclid's. He describes numbers and presents some concepts, makes excursions into philosophy, but never proves anything. In strong contrast to Euclid, who mentions not one specific number in his arithmetical books, he always uses numerical examples. A few selected quotations will bring out the flavor of Nicomachus's work. They are all from the translation of M. D'Ooge (1926).

After some general introductory remarks, Nicomachus mentions arithmetic, geometry, music (harmonics), and astronomy. He then argues that arithmetic is the foremost of these:

I, Ch. IV:

[1] Which then of these four methods must we first learn? Evidently, the one which naturally exists before them all, is superior and takes the place of origin and root and, as it were, of mother to the others.

[2] And this is arithmetic, not solely because we said that it existed before all the others in the mind of the creating God like some universal and exemplary plan, relying upon which as a design and archetypal example the creator of the universe sets in order his material creations and makes them attain to their proper ends; but also because it is naturally prior in birth, inasmuch as it abolishes other sciences with itself, but is not abolished together with them.

I, Ch. V:

[3] So then we have rightly undertaken first the systematic treatment of this, as the science naturally prior, more honorable, and more venerable, and, as it were, mother and nurse of the rest; and here we will take our start for the sake of clearness.

Only in Ch. VII does Nicomachus start to speak about numbers:

I, Ch. VII:

[1] Number is limited multitude or a combination of units or a flow of quantity made up of units; and the first division of number is even and odd.

[2] The even is that which can be divided into two equal parts without a unit intervening in the middle; and the odd is that which cannot be divided into two equal parts because of the aforesaid intervention of a unit.

[3] Now this is the definition after the ordinary conception; by the Pythagorean doctrine, however, the even number is that which admits of division into the greatest and the smallest parts at the same operation, greatest in size and smallest in quantity, in

accordance with the natural contrariety of these two genera; and the odd is that which does not allow this to be done to it, but is divided into two unequal parts.

In his definition of number, Nicomachus goes beyond Euclid in allowing "a flow of quantity made up of units" to be a number, for instance days or years or, we might add, segments built up from unit segments.

Prime numbers are introduced in Ch. XI.

I, Ch. XI:

[2] Now the first species, the prime and incomposite, is found whenever an odd number admits of no other factor save the one with the number itself as denominator, which is always unity; for example, 3, 5, 7, 11, 13, 17, 19, 23, 29, 31. None of these numbers will by any chance be found to have a fractional part with a denominator different from the number itself, but only the one with this as denominator, and this part will be unity in each case; for 3 has only a third part, which has the same denominator as the number and is of course unity, 5 a fifth, 7 a seventh, and 11 only an eleventh part, and in all of them these parts are unity.

He goes on to describe relatively prime pairs of numbers, the sieve of Eratosthenes for the identification of prime numbers, and similar things. One topic that has become popular outside of mathematics, even if only because of the name, is that of perfect numbers. Euclid merely shows how to construct perfect numbers in his Book IX, but Nicomachus describes them:

I, Ch. XVI:

[2] Now when a number, comparing with itself the sum and combination of all the factors whose presence it will admit, neither exceeds them in multitude nor is exceeded by them, then such a number is properly said to be perfect, as one which is equal to its own parts. Such numbers are 6 and 28; for 6 has the factors half, third, and sixth, 3, 2, and 1, respectively, and these added together make 6 and are equal to the original number, and neither more nor less. Twenty–eight has the factors half, fourth, seventh, fourteenth, and twenty–eighth, which are 14, 7, 4, 2 and 1; these added together make 28, and so neither are the parts greater than

the whole nor the whole greater than the parts, but their comparison is in equality, which is the peculiar quality of the perfect number.

[5] It comes about that even as fair and excellent things are few and easily enumerated while ugly and evil ones are widespread, so also the superabundant and deficient numbers are found in great multitude and irregularly placed—for the method of their discovery is irregular—but the perfect numbers are easily enumerated and arranged with suitable order; for only one is found among the units, 6, only one other among the tens, 28, and a third in the rank of the hundreds, 496 alone, and a fourth within the limits of the thousands, that is, below ten thousand, 8128. And it is their accompanying characteristic to end alternately in 6 or 8, and always to be even.

Superabundant numbers are those the sum whose of factors (parts) is greater than the number itself, for deficient numbers the sum is less.

Some of Nicomachus's topics are quite elementary. In I, Ch. XIX, for instance, he presents a multiplication table. Before doing this, however, he allows us to have a look into a classroom in antiquity:

I, Ch. XVII:
[6] The unequal, on the other hand, is split up by subdivisions, and one part of it is the greater, the other the less, which have opposite names and are antithetical to one another in their quantity and relation. For the greater is greater than some other thing, and the less again is less than another thing in comparison, and their names are not the same, but they each have different ones, for example, "father" and "son," "striker" and "struck," "teacher" and "pupil," and the like.

Part of Nicomachus's Book II is devoted to the geometric representation of numbers. As a sample we take his description of square numbers (after the triangular numbers):

II, Ch. IX:
[1] The square is the next number after this, which shows us no longer 3, like the former, but 4 angles in its graphic representation,

but is none the less equilateral. Take, for example, 1, 4, 9, 16, 25, 36, 49, 64, 81, 100; for the representations of these numbers are equilateral, square figures, as here shown; and it will be similar as far as you wish to go:

$$
\begin{array}{ccccc}
 & & & & \alpha\ \alpha\ \alpha\ \alpha\ \alpha \\
 & & & \alpha\ \alpha\ \alpha\ \alpha & \alpha\ \alpha\ \alpha\ \alpha\ \alpha \\
 & & \alpha\ \alpha\ \alpha & \alpha\ \alpha\ \alpha\ \alpha & \alpha\ \alpha\ \alpha\ \alpha\ \alpha \\
 & \alpha\ \alpha & \alpha\ \alpha\ \alpha & \alpha\ \alpha\ \alpha\ \alpha & \alpha\ \alpha\ \alpha\ \alpha\ \alpha \\
\alpha & \alpha\ \alpha & \alpha\ \alpha\ \alpha & \alpha\ \alpha\ \alpha\ \alpha & \alpha\ \alpha\ \alpha\ \alpha\ \alpha \\
1 & 4 & 9 & 16 & 25
\end{array}
$$

[2] It is true of these numbers, as it was also of the preceding, that the advance in their sides progresses with the natural series. The side of the square potentially first, 1, is 1; that of 4, the first in actuality, 2; that of 9, actually the second, 3; that of 16, the next actually the third, 4; that of the fourth, 5; of the fifth, 6, and so on in general with all that follow.

[3] This number also is produced if the natural series is extended in a line, increasing by 1, and no longer the successive numbers are added to the numbers in order, as was shown before, but rather all those in alternate places, that is, the odd numbers. For the first, 1, is potentially the first square; the second, 1 plus 3, is the first in actuality; the third, 1 plus 3 plus 5, is the second in actuality; the fourth, 1 plus 3 plus 5 plus 7, is the third in actuality; the next is produced by adding 9 to the former numbers, the next by the addition of 11, and so on.

Nicomachus's work seems to have been much more popular in antiquity than Euclid's *Elements*. Many more manuscripts of the *Introduction to Arithmetic* have survived than of the *Elements*.

Diophantus

Diophantus of Alexandria may have lived around 250 C.E.. His main work is the *Arithmetica* consisting of thirteen books, six of which survive in their original form and four more in Arabic translations. The first commentator on Diophantus was Hypatia, the daughter of Theon of Alexandria, around 400 C.E.. Diophantus's style is again very different from either Euclid's or Nicomachus's. He poses problem after problem, always with solutions in specific numbers, and no connecting text, no theory, no comment. By stating suitable conditions, he makes sure to get a rational solution for any of his problems. From this we have today's custom to speak of Diophantine problems if we will accept only integer (or rational) solutions.

In what follows we quote the introductory remarks of Diophantus, some elementary problems from his Book I, and one example from a later book. This last one is typical of his advanced problems and is presented in order to show how skillfully Diophantus manipulates very big numbers. (Our text is based on the English translation by Heath and the German one by Czwalina. For a short but very instructive introduction to Diophantus's work see Stillwell [1998], Ch. 4.)

18.2 Diophantus: The *Arithmetica*

From Book I:

> Dedication.
> Knowing, my most esteemed friend Dionysius, that you are anxious to learn how to investigate problems in numbers, I have tried, beginning from the foundations on which the science is built up, to set forth to you the nature and power subsisting in numbers.
>
> Perhaps the subject will appear rather difficult, inasmuch as it is not yet familiar (beginners are, as a rule, too ready to despair of success); but you, with the impulse of your enthusiasm and the benefit of my teaching, will find it easy to master; for eagerness to learn, when seconded by instruction, ensures rapid progress.
>
> . . .
>
> **Problem 1.** *To divide a given number into two having a given difference.*
> Given number 100, given difference 40.
> Lesser number required x. Therefore
>
> $$2x + 40 = 100,$$
> $$x = 30.$$

The required numbers are 70, 30.

Problem 2. *To divide a given number into two having a given ratio.*

Given number 60, given ratio $3 : 1$.
Two numbers x, $3x$. Therefore $x = 15$.
The numbers are 45, 15.

. . .

Problem 27. *To find two numbers such that their sum and product are given numbers.*

Necessary condition. The square of half the sum must exceed the product by a square number.

Given sum 20, given product 96.
$2x$ the difference of the required numbers.
Therefore the numbers are $10 + x$, $10 - x$.
Hence $100 - x^2 = 96$.
Therefore $x = 2$, and
the required numbers are 12, 8.

For this problem we may offer a geometric and algebraic interpretation. We are given a rectangle with sides y, z and know the sum $b = y + z$ of the sides and the area $c = yz$. Note that because Diophantus deals with numbers, he has no difficulties with dimensions, as did Euclid, who would see c as a rectangle in contrast to the segments y, z, b.

A modern solution would use $z = b - y$ and have

$$
\begin{aligned}
y(b - y) &= c, \\
y^2 - by + c &= 0, \\
\left(y - \frac{b}{2}\right)^2 &= \frac{b^2}{2} - c, \\
y &= \frac{b}{2} + \sqrt{\frac{b^2}{4} - c}, \\
z &= \frac{b}{2} - \sqrt{\frac{b^2}{4} - c}.
\end{aligned}
$$

Diophantus's solution is equally general. He has $y - z = 2x$ and

$$y = \frac{b}{2} + x \quad \text{and} \quad z = \frac{b}{2} - x,$$

$$c = \frac{b^2}{4} - x^2,$$

$$x^2 = \frac{b^2}{4} - c.$$

Thus he will get y, z as above. His condition "$\frac{b^2}{4} - c$ must be a square" makes sure that he has a rational solution. (This is similar to Euclid's condition in Prop. VI.28.)

The inherent structure of a quadratic problem is always the same, independent of the specific ways of literary expression.

From Book V:

Preceding the problem quoted below are some lemmas. The first one (to Probl. V. 6) shows how to find different right triangles with equal area. Diophantus presents the triangles with sides (40, 42, 58), (24, 70, 74), (15, 112, 113), the area of each being 840. The next lemma shows how to find three numbers such that the products of the three pairs from these numbers shall be respectively equal to three given squares.

Problem 8. *To find three numbers a, b, c such that the product of any two of them plus or minus their sum $a + b + c$ gives a square.*

Solution: As in the first lemma, we find three right angled triangles with equal areas; the squares of their hypotenuses are 3364, 5476, 12769. Now find, as in the second lemma, three numbers p, q, r such that the products of the three pairs are equal to these squares respectively.

We took the above numbers because each of them plus or minus the fourfold area, which is 3360, gives a square.

The three numbers p, q, r are then

$$p = \frac{4292}{113}, \quad q = \frac{380132}{4292}, \quad r = \frac{618788}{4292},$$

The products of any two of these are the above squares.

We will need to have the sum of the three numbers equal to $3360x^2$. For that, calculate, with the common denominator 484996, for p, q, r:

$$px + qx + rx = \frac{1}{484996}(18421264 + 42954916 + 69923044)x$$

$$= \frac{131299224}{48499x}x.$$

With this equal to $3360x^2$ we find

$$x = \frac{131299224}{1629586560} = \frac{781543}{9699920}.$$

Hence the three numbers are

$$a = \frac{781543}{255380}, \quad b = \frac{781543}{109520}, \quad c = \frac{781543}{67280}.$$

(The program Derive confirms Diophantus's assertions. For instance,

$$bc - (a + b + c) = \left[\frac{75809671}{9699920}\right]^2.$$

What Diophantus could not find was a rational right triangle with square area. Fermat showed that such a triangle does not exist; see Stillwell [1998].)

19

CHAPTER

Euclid Book VIII

Numbers in Continued Proportion, the Geometry of Numbers

19.1 The Overall Composition of Books VIII and IX

There is no break in the contents between Books VIII and IX. Book IX just continues to treat the problems from the end of Book VIII. For that reason we will show the internal subdivisions of the two books in one picture.

Book VIII	
1–3	A: Numbers in continued proportion
4–5	Two singletons
6–10	A: continued
11–27	B: The geometry of numbers: similar plane and solid numbers, mean proportionals

(continued)

19.2 Book VIII, Part A: Numbers in Continued Proportion

When Euclid speaks of "as many numbers as we please in continued proportion," as usual he takes a few numbers a, b, c, d and assumes

$$a : b = b : c = c : d,$$

which in today's formulas could be expressed (with $b = aq$ etc.) as

$$a : aq = aq : aq^2 = aq^2 : aq^3,$$

so that we are speaking of a geometric sequence like a, aq, aq^2, aq^3. There are certain inhomogeneities in Euclid's text that are discussed by various authors; see; e.g., Mueller [1981, p. 83.].

Propositions VIII.1–3 generalize the properties of minimi from Props. VII.21/22 to numbers in continued proportion. Proposition VIII.5 is a singleton closely related to Prop. VI.23, which is similarly isolated in Book VI.

Prop. VIII.5.
Plane numbers have to one another the ratio compounded of the ratios of their sides.

Let the plane numbers be $a = cd$ and $b = ef$ with their "sides" c, d and e, f respectively. Then the ratio $a : b$ is "compounded" from $c : e$ and $d : f$. (There is no formal definition of compounding in the *Elements*.) In the modern notation of fractions this is simply

$$\frac{a}{b} = \frac{cd}{ef} = \frac{c}{e} \cdot \frac{d}{f}.$$

This is as close as Euclid ever comes to any operation with fractions. In particular, there is no trace whatsoever of the addition of fractions or ratios. Ratios are not numbers for Euclid. We have seen that this is very different for Diophantus.

In Prop. VI.23 practically the same statement as in VIII.5 is made for rectangles (or equiangular parallelograms) and their sides. The proof of VIII.5 follows exactly the lines of VI.23. Again, the arithmetical proof uses VII.17 where the geometrical one has VI.1. Both VI.23 and VIII.5 are never used again in the *Elements*. They look like later additions to the text.

After the interpolation of Props. VIII. 4/5 the text goes on about numbers in continued proportion. As usual, Euclid gives no numerical example, but we may as well do so and look at

$$81 : 54 \quad = \quad 54 : 36 \quad = \quad 36 : 24 \quad = \quad 24 : 16,$$
$$p^4 : (p^3 q) \quad = \quad (p^3 q) : (p^2 q^2) \quad = \quad (p^2 q^2) : (pq^3) \quad = \quad (pq^3) : q^4.$$

The logical chain of the propositions in part A of Book VIII is straightforward: $(1) \Rightarrow (2) \Rightarrow (3) \Rightarrow (6) \Rightarrow (7)$. Throughout, heavy use is made of the theory of minimi from Book VII. We have seen that this is Euclid's substitute for the uniqueness of prime factorization.

Prop. VIII.7.
If there are as many numbers as we please in continued proportion, and the first measures the last, it will measure the second also.

This has important implications in part B.

19.3   Book VIII, Part B: The Geometry of Numbers

Plato writes about proportions and means:

> Two things cannot be rightly put together without a third; there must be some bond of union between them. And the fairest bond is that which makes the most complete fusion of itself and the things which it combines, and proportion is best adapted to effect such a union. For whenever in any three numbers, whether cube or square, there is a mean, which is to the last term what the first term is to it. . . .
>
> If the universal frame had been created a surface only and having no depth, a single mean would have sufficed to bind together itself and the other terms, but now, as the world must be a solid, and solid bodies are always compacted not by one mean but by two. . . . (Timaeus 31b–32b)

These lines have to be read in the context of Plato's cosmology, but for us it is enough to see how important the mean proportionals have been for him. He postulates one mean proportional in the plane case and two mean proportionals in the solid case. Moreover, he speaks of numbers, but combines them with aspects of surfaces and solids. We find this same combination and terminology in Euclid's Book VIII. Here again the proofs are purely arithmetical, but the leading ideas come from geometry, as we will see.

Euclid gives the plane and the solid cases a parallel treatment. For instance, Props. 14, 18, 20, 16 are about "plane" numbers, and Props. 15, 19, 21, 27 are about "solids." The solid case has close connections to some theorems about solid bodies in Book XI. It turns out that the question of two mean proportionals to numbers (or segments) a, b, like $a : x = x : y = y : b$, is closely related to the problem of doubling a cube. That is why we will postpone the solid case to our study of Book XI and content ourselves here with "plane" numbers.

The problem of finding a mean proportional m for any segments a, b such that $a : m = m : b$ has been solved in Prop. VI.13. The solution is general. Whatever the segments a, b are, it is always possible to construct m, essentially with the help of the theorem of

Pythagoras. The arithmetical problem is very different: For some pairs of numbers, like $a = 3$ and $b = 12$, a mean proportional $m = 6$ exists, but for others, like $a = 4$ and $b = 12$ it does not exist. Hence we have the attractive problem of characterizing those pairs a, b of numbers that have a mean proportional m. Euclid solves it in Props. VIII, 18/20, basically by introducing the concept of "similar plane numbers" as defined in Def. VII.16. The whole group of Defs. VII.16–19 belongs to this section of Book VIII. The notions of duplicate and triplicate ratios, which are used in this context, are defined in Book V, Defs. 9 and 10. Magnitudes (and numbers, we may assume) a, b are in the duplicate ratio of $a : m$ if $a : m = m : b$, and similarly for the triplicate ratio with two means. No use is made of compound ratios in this context.

A word more should be said about similar plane numbers. Similar plane numbers are those that have their sides proportional. It remains an open question whether Euclid would accept $1 \cdot 5 = 5$ and $1 \cdot 5 = 5$ as similar plane numbers. (Remember that Plato called 5 a rectangular number.) Cases like $a = 2 \cdot 3 = 6$ and $b = 2 \cdot 3 = 6$ as plane numbers and $m = 6$ as mean proportional would certainly be of no great interest for Euclid, but they are not excluded. However, in the proof of Prop. VIII.20 it would turn out, in this case, that $1 : 1$ would be the minimi for $6 : 6$, and $1 \cdot 6$ should be considered to be plane. The best we can do is to leave these difficulties aside and always look at the "general" case. We will follow Euclid's text step by step, but ignore cubes and less essential remarks. As a shorthand, we will use modern notation, and occasionally write $x|y$ for "x measures (or divides) y."

Prop. VIII.11.
Between two square numbers $a = c^2$ and $b = d^2$ there is one mean proportional number cd, and the square has to the square the ratio duplicate of that which the side has to the side.

This is the easy first case of the general problem with the easy proof for $c^2 : cd = cd : d^2$. (Remember from the introduction to Book V that the Parthenon temple in Athens is constructed according to the ratio $9 : 4$ throughout.)

Prop. VIII.14.
$a^2|b^2 \Leftrightarrow a|b$.

The proof of this proposition, especially for the case $a^2|b^2 \Rightarrow a|b$, relies upon Prop. VIII.7 and hence, as we have mentioned, on Euclid's substitute "minimi" for the unique factorization of primes.

Prop. VIII.18.
(i) *Between two similar plane numbers there is one mean proportional number.*

Prop. VIII.20.
If one mean proportional number falls between two numbers, the numbers will be similar plane numbers.

Prop. VIII.18
continued: (ii) *And the plane number has to the plane number the ratio duplicate of that which the corresponding side has to the corresponding side.*

Proof of 18 (i).
Let a, b be the two similar plane numbers with sides c, d of a and e, f of b, and $c : d = e : f$ and $a = cd$ and $b = ef$ according to the definition of similar plane numbers. For the following arguments using a gnomon, see Fig. 22. Euclid shows only segments as in Fig. 17.4 (a):

$$c : d = e : f \Rightarrow c : e = d : f \qquad \text{by alternation.}$$

Let $g = ed$ an auxiliary number. Then

$$a : g = cd : ed = c : e = d : f,$$
$$g : b = ed : ef = d : f;$$

hence $a : g = g : b$ and a, b have the mean proportional g.

Proof of Prop. 20:
Assume $a : m = m : b$. It has to be shown that a, b are plane numbers and that their sides are proportional. Let $d : e$ be minimi for $a : m$, and hence for $m : b$ as well. Then there are numbers k, n such that $a = kd$ and $m = ke$ and $m = nd$ and $b = ne$. This is a factorization for a and b, hence, they are plane numbers. We have to show that

FIGURE 19.1 $a = 6 \cdot 8$ and $b = 9 \cdot 12$ and $6 : 8 = 9 : 12$ and $m = 72 = 8 \cdot 9 = 6 \cdot 12$.

their sides k, d and n, e are proportional. We know that

$$d : e = (kd) : (ke) = a : m = m : b,$$
$$m : b = (nd) : (ne) = (ke) : (ne)$$
$$= k : n$$

because $nd = m = ke$.

From $d : e = k : n$ we derive by alternation

$$d : k = e : n,$$

and the sides of a and b are proportional.

In spite of the same numbers, the picture of this proof looks different from Fig. 19.1. Assume $a = 48$ and $m = 72$ and $b = 108$. Then the minimi are $2 : 3 = d : e$, and we will have $k = 24$ and $n = 36$, and the picture will look something like Fig. 19.2.

Observe that nd and ke appear as the "so-called complements" of "the schema" of Prop. I.43.

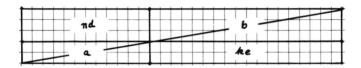

FIGURE 19.2

Proof of Prop. VIII.18 (ii):
The similar plane numbers are a, b as in 18 (i) with factors $a = cd$ and $b = ef$ and $c : d = e : f$. We had $g = ed = cf$ and $a : g = c : e = d : f$ and $a : g = g : b$. By definition, $a : b$ is the duplicate ratio of the ratio of the sides $c : e$.

Let us try to discuss the problem of Prop. VIII.20 once more, but this time using the modern tool of unique prime factorization. Euclid uses minimi and goes back through a whole series of preparatory statements. By establishing a clear mathematical picture, we will obtain a clearer view of the historical situation.

Let a, m, b be numbers such that

$$a : m = m : b \qquad \Leftrightarrow \qquad ab = m^2.$$

Assume that a has the prime factorization

$$a = r^2 p_1 \ldots p_t, \text{ where } \qquad p_1 \ldots p_t \qquad \text{is square-free.}$$

Then, because $ab = m^2$, the prime factorization of b must be

$$b = s^2 p_1 \ldots p_t,$$

with some square s^2 and $p_1 \ldots p_t$ as above. For brevity write $p_1 \ldots p_t = q$; hence $a = r^2 q$ and $b = s^2 q$ with q square–free.

Return from this point to the definition of similar numbers with sides c, d of a and e, f of b proportional and let x, y be minimi for $c : d$. Then $xy = q$ is the "area" of the small rectangle contained by the sides x, y. Because x, y are minimi, we find r, s such that $c = rx$, $d = ry$ and $e = sx$, $f = sy$. The small rectangle enlarged by the factor r produces the plane number a and enlarged by s produces the similar plane number b.

Before going on to the general situation let us isolate two extreme cases. If $a = r^2$ and $b = s^2$, we have

$$r^2 : rs = rs = s^2,$$

as in VIII.11.

If $a(= q)$ and b are square-free, then

$$q : q = q : q$$

is the only possibility because of the unique factorization into primes.

What makes Euclid's solution interesting is his combination of the extreme cases into a general solution of his problem. We return to $a = r^2q$ and $b = s^2q$. This implies $m = rsq$, again by unique prime factorization. Since a common factor of r, s could be canceled, let us assume that $gcd(r, s) = 1$ and r, s are the corresponding minimi d, e of Euclid. Then

$$a : m \;=\; r : s \quad \text{and} \quad a = rk, \quad m = sk \quad \text{with} \quad k = rq;$$
$$m : b \;=\; r : s \quad \text{and} \quad m = rn, \quad b = sn, \quad \text{with} \quad n = sq.$$

Hence the "sides" of the numbers a, b are r, k and s, n respectively. From the equality $rn = sk$ we derive

$$r : k = s : n,$$

and the "sides" of the "plane numbers" are proportional.

This modern analysis of similar plane numbers will suffice to make the next two statements of Euclid obvious. Euclid himself goes back to Prop. VIII.20.

Prop. VIII.26.

If a, b are similar plane numbers, then there exist numbers r, s such that

$$a : b = r^2 : s^2.$$

Prop. IX. 1, 2:

a, b are similar plane numbers $\Leftrightarrow$ There exists a number m such that $ab = m^2$.

20

The Origin of Mathematics 10

Tools and Theorems

··

The Greek word *theorema* is related to *theater* and means *"something seen,"* or, in mathematics, *an insight*, understanding, or knowledge. A mathematical theorem should answer a question or solve a problem that posed itself during the discussion of a mathematical subject. Very often, a theorem, once proved, is used in various ways to develop a mathematical theory further. In this process it changes its character. The original questions recede, and the theorem becomes a basic tool for the practitioners.

Before we go on to show how this general feature of mathematical theories is present in the *Elements*, a typical example from modern mathematics may serve to illustrate the situation.

One of the first propositions in the elementary theory of finite groups is Lagrange's theorem: *If G is a finite group of order g and S a subgroup of G of order s, then s divides g*. This is usually presented together with an example of a group *G* of order 12 that has no subgroup of order 6. Are there any special divisors *s* of *g* such that there must be a subgroup *S* of order *s* in *G*? The answer to this problem is a highlight at the end of the elementary part of group theory, namely

203

Sylow's theorem: *If* $s = p^{\alpha}$ *is a prime power dividing g, then there exists a subgroup S of order* p^{α} *in G* (plus some more information). As soon as one goes on into a little more advanced parts of the theory of finite groups, Sylow's theorem becomes a universal and indispensable tool that is used over and over again in the proofs of deeper theorems. This double character of theorems and tools is commonplace in mathematics; the most famous example may be the fundamental theorem of calculus about derivatives and integrals.

As in the case of Sylow's theorem, we have seen Euclid's theory of similar plane numbers providing the answer to the problem about mean proportionals for numbers. And just as the group theorists proceed with Sylow's theorem, Euclid uses similar plane numbers as a tool in his more advanced Book X. For certain purposes he needs what we call Pythagorean triples, integers x, y, z such that $x^2 + y^2 = z^2$.

Book X, Lemma 1 to Prop. X. 28.
To find two square numbers such that their sum is also a square.

Proof.
This time we follow Euclid's practice and present numbers by segments. Let AB, BC be similar plane numbers and let $AB > BC$ and let both of them be even or both odd. Then $AB - BC$ is even, and $AD = \frac{1}{2}(AB - BC)$ is an integer (Fig. 20.1).

Because D bisects AC, and CB is added, we may apply Prop. II.6 and get

$$AB \cdot BC + CD^2 = BD^2.$$

Moreover, because AB, BC are similar plane numbers, $AB \cdot BC = FE^2$ is a square by Prop. IX. 1. Hence we have found a Pythagorean triple.

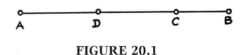

FIGURE 20.1

Comment

(Modern notation) If the two numbers a, b are both even or both odd, we will have $a - b = 2w$, with an integer w. Then $a = b + 2w$ and we have

$$ab + w^2 \;=\; (b + 2w)b + w^2 = (b + w)^2.$$

This was the application of Prop. II.6. If a, b are similar plane numbers, then $ab = m^2$, and we have the Pythagorean triple

$$m^2 + w^2 = (b + w)^2.$$

At this point we will introduce the factorizations of similar plane numbers found in Book VIII, Section 3:

$$a \;=\; r^2 q \text{ and } b \;=\; s^2 q \;\;; \text{hence } 2w = (r^2 - s^2)q.$$

$$m^2 \;=\; ab = r^2 s^2 q^2 \;=\; x^2.$$

$$w^2 \;=\; \left[\frac{r^2 - s^2}{2}\right]^2 q^2 \;=\; y^2.$$

$$(b + w)^2 \;=\; \left[\frac{r^2 + s^2}{2}\right]^2 q^2 \;=\; z^2.$$

These formulas are nothing else but the well-known parametrization of all Pythagorean triples. Euclid does not say that he has found all of them, but in fact he has. The modern proof for this assertion can be found in any book about number theory. See, for instance, Stillwell [1998] Ch. 4.

Two more general comments should be made. The first one is to observe that the numbers b and w are themselves "plane," like $2w = a - b = (r^2 - s^2)q$, and these numbers have again to be squared. For a square of a square there is no geometric representation. Hence at this point the subject matter itself forces Euclid to represent "plane" numbers by segments. The second remark concerns the application of the geometrical Prop. II.6 to numbers. We can see this in two ways. Either Euclid identifies numbers with segments, or he has a more abstract algebraic understanding of Prop. II.6. Because he uses geometric ideas for his proofs all the time, the first alternative seems to be more likely.

Two More Examples

Another typical example for the tools/theorems dual aspects is Prop. III.21/22, which we discussed earlier. Proposition III.22 solves the problem of characterizing quadrangles in circles, and Prop. III.21 establishes the tool of the invariance of angles in segments.

In Prop. VI.2 Euclid presents first the tool (the theorem on proportional segments), and afterwards the theorem (Props. VI.4/5 about similar triangles).

Despite its name, the theorem of Pythagoras, Prop. I.47, somehow lacks one important feature of a theorem. It does not answer a natural question, as do the theorems about mean proportionals and cyclic quadrangles. It is hard to see what the original questions leading to the theorem of Pythagoras might have been, and so far nobody has proposed a convincing hypothesis. Thus in the classification of this section, we have to regard Pythagoras's theorem more as a tool than as a theorem. But on the other hand, Pythagoras's theorem is so important exactly because it brings something to light that could *not* be anticipated.

21
CHAPTER

Euclid Book IX

Miscellaneous Topics from Arithmetic

21.1 Book IX, Parts B, C, D, and E: More About Numbers in Continued Proportion

The first two propositions of Book IX have already been dealt with in the section on similar plane numbers. Props. IX. 3–6 are concerned with cube numbers, for instance

Prop. IX. 3.
x is a cube $\Rightarrow x^2$ is a cube.

Prop. IX. 6.
x^2 is a cube $\Rightarrow x$ is a cube.

The singleton Prop. IX. 7 looks curious, because it seems just to repeat Def. VII.17 of a solid number:

Prop. IX. 7.

If a composite number by multiplying any number makes some number, the product will be solid.

The proof goes like this: Let $ab = c$ and a be composite, $a = de$. Then $c = (de)b$. By some sort of handwaving, Euclid deletes the (for him invisible) parentheses and has what he wants. We can offer the following explanation. Def. VII.17 takes it for granted that we can form a product of three factors. It seems to be older than the asymmetric Def. VII.15 of one number multiplying another one, which in a way excludes three factors. In fact, Def. VII.15 is quoted in the proof of IX. 7. It seems that the purpose of Prop. IX. 7 is to reconcile Definitions 5 and 17 from the beginning of Book VII.

Why did it find its place right here? We have already seen other similarly isolated items. It was a scribal practice in antiquity to put suitable additions to a text at the end of a papyrus roll if there was some space left. Maybe that is the way Prop. VI, 23, Prop. VIII.5 and Prop. IX, 7 came to the places where they are.

Propositions IX. 8–11 are about numbers in continued proportion starting with 1. Such a sequence would look in modern notation like

$$1, a, a^2, a^3, a^4, a^5, a^6, a^7, \ldots.$$

The propositions make some observations about squares and cubes in sequence that are obvious when seen in the modern way. Part D of Book IX has been discussed in Book VII, in the section about prime divisors. Part E is a sort of sequel to the theory of similar plane numbers from Book VIII. In Book VIII the problem was to find a mean proportional m for two numbers a, b such that $a : m = m : b$. This time the question concerns the existence of third and fourth proportionals.

(i) Let a, b be given. When and how can we find an integer x such that

$$a : b = b : x \,?$$

(ii) Let a, b, c be given. When and how can we find an integer x such that

$$a : b = c : x ?$$

Euclid's answers are rather involved and are of no special interest to us.

21.2   Book IX, Part F: The Number of Primes

This section consists of just one theorem and its proof. Both together are a jewel of mathematics.

Prop. IX. 20. The theorem about the number of primes.
There are more prime numbers than any assigned multitude of prime numbers.

Proof.
(We abbreviate very little). Let a, b, c be the assigned prime numbers; I say that there are more prime numbers than a, b, c. For consider $d = abc + 1$ (the least common multiple abc and add 1). It is either prime or has a prime factor g (by Prop. VII.32). If d is prime, then the prime numbers a, b, c, d have been found which are more than a, b, c.

If d is not prime but has the prime factor g, then I say that g is not the same with any of the numbers a, b, c.

For, if possible, let it be so. Now since a, b, c divide abc, g would also divide abc. But it divides $d = abc + 1$. Therefore g, being a number, would divide the remainder, the unit 1: which is absurd. Therefore g is not equal to any one of a, b, c. And by hypothesis it is prime.

Therefore the prime numbers a, b, c, g have been found which are more than the assigned multitude of a, b, c.

We will return to this proof in the next chapter on the beauty of mathematics. For the moment let us look at numerical examples, about which Euclid, as usual, keeps quiet.

$$2 \cdot 3 \cdot 5 \cdot 7 + 1 \quad = \quad 211 \qquad \text{is prime;}$$
$$2 \cdot 3 \cdot 5 \cdot 7 \cdot 11 \cdot 13 + 1 \quad = \quad 30031 \; = \; 59 \cdot 509 \qquad \text{is not prime}$$
$$\text{but has the (new) prime factor 59.}$$

21.3   Book IX, Part G: Odd and Even Numbers, and Perfect Numbers

Plato writes in his dialogue *Gorgias*:

> Suppose that somebody should ask me ... : Socrates, what is the art of arithmetic? I should reply ... that it is one of the arts which secure their effect through speech. And if he should further inquire in what field, I should reply that of the odd and the even, however great their respective numbers might be. (*Gorgias* 451 ab)

In Euclid's Book IX, Defs. 6–10 and Props. 21–33, we find exactly what Plato describes, the theory of odd and even numbers. Mathematically, it is not overly exciting; see for instance

Prop. IX. 25.
If from an even number an odd number is subtracted, the remainder will be odd.

Historically, the piece is interesting because it is obviously a self-contained little theory that was known to Plato, who mentions it not only in the quotation above but at several other places as well (see Notes). O. Becker [1934] was the first to understand it as a piece of pre–Euclidean mathematics that was preserved in its entirety in the *Elements*. With this pioneering work Becker started a whole direction of historical research that tries to find traces of pre–Euclidean theories in Euclid's writings.

Perfect Numbers

Euclid has defined (Def. VII.22) a perfect number to be one that is equal to its own parts. From Nicomachus we know that

$$496 = (1 + 2 + 4 + 8 + 16) + (31 + 62 + 124 + 248)$$

is perfect. (1 is considered to be a part of 496, but not 496 itself.) Euclid sets out to prove a general theorem:

Prop. IX. 36.
If the sum $1 + 2 + \ldots + 2^k = p$ is prime, then the number $p2^k$ will be perfect.

For the proof three things have to be known:
  (i) $1, 2, \ldots, 2^k$ and $p, 2p, \ldots, 2^{k-1}p$ are "parts," i.e. divisors, of $p2^k$. This is easy.
 (ii) There are no other divisors of $p2^k$ than the ones from (i). This is harder to do without prime factorization. Euclid uses the *minimi* from Prop. VII.20 as a substitute for prime factorization in the proof.
(iii) There has to be a way to find the sum of $1 + 2 + 4 + \ldots + 2^k$ and of $p(1 + 2 + \ldots + 2^{k-1})$. This is the subject of Prop. IX. 35.

If we are allowed to employ prime factorization and modern formulas, all three parts are easily done. Euclid's cumbersome notation makes things more difficult, but his procedure is essentially the modern one. The interested reader may look up Euclid's original proof.

The Origin of Mathematics
11

Math Is Beautiful

....

When mathematicians praise the aesthetic qualities of their subject, they usually meet raised eyebrows from sceptical nonbelievers. What will be said here will probably not convince outsiders, but it may clarify the situation somewhat for ourselves. Using examples as well as comments by some prominent authors, we will try to be more specific than just stating an emotional opinion. Proclus will again be our main witness from antiquity. He discusses the applications of mathematics and then goes on to describe its superior beauty and its value for the study of philosophy. In part he elaborates a short passage from Aristotle, who says what beauty is: "The chief forms of beauty are order and symmetry and definiteness, which the mathematical sciences demonstrate in a special degree" (Aristotle, *Metaphysics* 1078 a34–b2). (Here, as in the quotation from Proclus below and in many other places, the translator has used the modern "symmetry" for the Greek "symmetria," which, at least in a mathematical context and by its literal translation, should mean "of common measure." In most cases the literal translation makes much more sense than the modern word "symmetry," which has taken on quite a different meaning outside of mathematics.)

About mathematics in general Proclus writes:

> There are nevertheless contentious persons who endeavor to detract from the worth of this science, some denying its beauty and excellence on the ground that its discourses say nothing about such matters, others declaring that the empirical sciences concerned with sense objects are more useful than the general theorems of mathematics. Mensuration, they say, is more than geometry, popular arithmetic than the theory of numbers, and navigation than general astronomy. For we do not become rich by knowing what wealth is but by using it, nor happy by knowing what happiness is but by living happily. Hence we shall agree, they say, that the empirical sciences, not the theories of the mathematicians, contribute most to human life and conduct. Those who are ignorant of principles but practised in dealing with particular problems are far and away superior in meeting human needs to those who have spent their time in the schools pursuing theory alone.
>
> To those who say these things we can reply by exhibiting the beauty of mathematics on the principles by which Aristotle attempts to persuade us. Three things, he says, are especially conducive to beauty of body or soul: order, symmetry, and definiteness. Ugliness in the body arises from the ascendancy of disorder and from a lack of shapeliness, symmetry, and outline in the material part of our composite nature; ugliness of mind comes from unreason, moving in an irregular and disorderly fashion, out of harmony with reason and unwilling to accept the principles it imposes; beauty, therefore, will reside in the opposites of these, namely, order, symmetry, and definiteness. These characters we find preeminently in mathematical science. We see order in its procedure of explaining the derivative and more complex theorems from the primary and simpler ones; for in mathematics later propositions are always dependent on their predecessors, and some are counted as starting points, others as deductions from the primary hypotheses. We see symmetry in the accord of the demonstrations with one another and in their common reference back to Nous; for the measure common to all parts of the science is Nous, from which it gets its principles and to which it directs

the minds of its students. And we see definiteness in the fixity and certainty of its ideas; for the objects of mathematical knowledge do not appear now in one guise and now in another, like the objects of perception or opinion, but always present themselves as the same, made definite by intelligible forms. If, then, these are the factors especially productive of beauty, and mathematics is characterized by them, it is clear that there is beauty in it.... We do not think it proper, moreover, to measure its utility by looking to human needs and making necessity our chief concern.... Just as we judge the usefulness or uselessness of the cathartic virtues in general by looking not to the needs of living, but rather to the life of contemplation, so we must refer the purpose of mathematics to intellectual insight and the consummation of wisdom. For this reason the cultivation of it is worthy of earnest endeavor both for its own sake and for the sake of the intellectual life. Evidence that it is intrinsically desirable to those who are engaged in it is, as Aristotle somewhere says, the great progress that mathematical science has made in a short time, although no reward is offered to those who pursue it, and the fact that even those who gain but slight benefit from it are fond of it and occupy themselves with it to the neglect of other concerns. So those who despise mathematical knowledge are they that have no taste for the pleasures it affords.

Consequently instead of crying down mathematics for the reason that it contributes nothing to human needs—for in its lowest applications, where it works in company with material things, it does aim at serving such needs—we should, on the contrary, esteem it highly because it is above material needs and has its good in itself alone. (Proclus–Morrow pp. 22–24)

Proclus strongly favors abstract mathematics because it purifies and elevates the soul, like Homer's Athena dispersing the mist (from Odysseus' homeland) obscuring the intellectual light of understanding (Proclus–Morrow, l.c. p. 25) What he says of beauty is equally general and philosophical, but in no way does it concern itself with any definite mathematical theorem or theory.

Modern authors, and especially G. H. Hardy [1967], whom we will follow for the next pages, have been more concrete and confined themselves to mathematics proper. (The quotations are from pages 84–115.) We, like Hardy, will not follow Aristotle and Proclus and try to define mathematical beauty. If we cannot give a definition of mathematical beauty, then that is just as true of beauty of any kind. We may not know what we mean by a beautiful poem or picture, but that does not prevent us from recognizing one when we read or see it. Even with poems or pictures a certain training or familiarity is necessary for any appreciation, and more than elsewhere in mathematics a minimal degree of understanding the tools of the trade is indispensable.

Hardy starts with some general introductory remarks:

> A mathematician, like a painter or a poet, is a maker of patterns. If his patterns are more permanent than theirs, it is because they are made with *ideas*. . . .
>
> The mathematician's patterns, like the painter's or the poet's, must be *beautiful*; the ideas, like the colours or the words, must fit together in a harmonious way. Beauty is the first test: there is no permanent place in the world for ugly mathematics. . . .
>
> The best mathematics is *serious* as well as beautiful—"important" if you like, but the word is very ambiguous, and "serious" expresses what I mean much better. . . . The "seriousness" of a mathematical theorem lies, not in its practical consequences, which are usually negligible, but in the *significance* of the mathematical ideas which it connects. We may say, roughly, that a mathematical idea is "significant" if it can be connected, in a natural and illuminating way, with a large complex of other mathematical ideas. Thus a serious mathematical theorem, a theorem which connects significant ideas, is likely to lead to important advances in mathematics itself and even in other sciences. . . .
>
> It will be clear by now that, if we are to have any chance of making progress, I must produce examples of "real" mathematical theorems, theorems which every mathematician will admit to be first-rate. . . . I can hardly do better than go back to the Greeks. I will state and prove two of the famous theorems of Greek mathematics. They are "simple" theorems, simple both in idea and in execution, but there is no doubt at all about their being theorems of the highest class. Each is as fresh and significant as when it was discovered—two thousand years have not written a wrinkle on either of them. . . . The first is Euclid's proof of the existence of an

infinity of prime numbers. . . . My second example is Pythagoras's proof of the irrationality of $\sqrt{2}$.

Hardy goes on to state and prove these two theorems. We have seen the first one in Prop. IX, 20 and will discuss the second one in detail in the section on incommensurability and irrationality. Then Hardy analyzes in more detail what he means by seriousness, significance, generality, and depth of a theorem. (It should be kept in mind that he means the irrationality of $\sqrt{2}$ when he speaks here of Pythagoras's theorem.) After that he returns to the subject of beauty in mathematics.

> What "purely aesthetic" qualities can we distinguish in such theorems as Euclid's and Pythagoras's? I will not risk more than a few disjointed remarks.
>
> In both theorems (and in the theorems, of course, I include the proofs) there is a very high degree of *unexpectedness*, combined with *inevitability* and *economy*. The arguments take so odd and surprising a form; the weapons used seem so childishly simple when compared with the far-reaching results; but there is no escape from the conclusions. There are no complications of detail—one line of attack is enough in each case; and this is true too of the proofs of many much more difficult theorems, the full appreciation of which demands quite a high degree of technical proficiency.

We could test many theorems from Euclid's *Elements* against these criteria, but with the exception of one example below, we leave this to the reader. At the end of his discussion of mathematical beauty, Hardy summarizes all his arguments in one powerful and compelling statement:

> It will be obvious by now that I am interested in mathematics only as a creative art.

One might not be too surprised to see such a confession from a number theorist like Hardy. It is really startling to read the same opinion from one of the most eminent mathematicians of our century. In a paper with the title "The Mathematician" John von Neumann, one of the inventors of the electronic computer, writes:

The mathematician has a wide variety of fields to which he may turn, and he enjoys a very considerable freedom in what he does with them. To come to the decisive point: I think that it is correct to say that his criteria of selection, and also those of success, are mainly aesthetical. I realize that this assertion is controversial and that it is impossible to "prove" it, or indeed to go very far in substantiating it, without analyzing numerous specific, technical instances. This would again require a highly technical type of discussion, for which this is not the proper occasion. Suffice it to say that the aesthetical character is even more prominent than in the instance I mentioned above in the case of theoretical physics. One expects a mathematical theorem or a mathematical theory not only to describe and to classify in a simple and elegant way numerous and a priori disparate special cases. One also expects "elegance" in its "architectural," structural makeup. Ease in stating the problem, great difficulty in getting hold of it and in all attempts at approaching it, then again some very surprising twist by which the approach, or some part of the approach, becomes easy, etc. Also, if the deductions are lengthy or complicated, there should be some simple general principle involved, which "explains" the complications and detours, reduces the apparent arbitrariness to a few simple guiding motivations, etc. These criteria are clearly those of any creative art, and the existence of some underlying empirical, worldly motif in the background – often in a very remote background, overgrown by aestheticizing developments and followed into a multitude of labyrinthine variants—all this is much more akin to the atmosphere of art pure and simple than to that of the empirical sciences.

You will note that I have not even mentioned a comparison of mathematics with the experimental or with the descriptive sciences. Here the differences of method and of the general atmosphere are too obvious.

I think that it is a relatively good approximation to truth— which is much too complicated to allow anything but approximations—that mathematical ideas originate in empirics, although the genealogy is sometimes long and obscure. But, once they are so conceived, the subject begins to live a peculiar life of its own and is better compared to a creative one, governed by almost entirely aesthetical motivations, than to anything else and, in particular, to an empirical science. (v. Neumann, p. 8/9)

In von Neumann's paper, geometry features prominently as a science originating in empirics. Here again, Euclid was the pioneer. We have seen, for instance, the transformation of the empirical science of "land measurement" into a mathematical theory in Euclid's Books I and II, with its culmination in the theorem Prop. II.14, which tells us how to "square" any polygonal area. Looking back and using von Neumann's words, we remember the elegance of the architecture, the two decisive steps in I.44 (simple application of areas) and I.47 (theorem of Pythagoras); and then its own peculiar life in Book II, where the solution is found with special tricks in order to avoid any similarity arguments. All this has been followed into a multitude of labyrinthine variants of measuring areas und volumes by a plethora of modern theories. On the other hand, the rampant growth of elementary geometry under the hands of French and German gymnasium teachers of the nineteenth century has led nowhere, and the results that filled journals are now largely forgotten.

Let us now return to Hardy's criteria for a beautiful proof: it should be *unexpected*, *inevitable*, and *economical*. There is a little gem in Euclid's Book IV that meets these conditions in an exceptionally fortunate way.

Prop. IV. 2.
In a given circle to inscribe a triangle equiangular with a given triangle.

Let $\mathcal{K}$ be the given circle (with radius R) and $\triangle DEF$ the given triangle with angles α, $\beta\gamma$ as shown in Fig. 22.1.

There are various ways to solve this problem. For instance, one could draw the circumcircle with radius r about $\triangle DEF$ and then enlarge (or reduce) $\triangle DEF$ in the ratio $R : r$. Instead of this pedestrian way, Euclid applies a very clever but simple idea.

Select any point A on $\mathcal{K}$ and draw the tangent GH to $\mathcal{K}$ in A (Fig. 22.2). Construct the angles $\angle GAC$ equal to β and $\angle HAB$ equal to γ. Then, by Prop. III. 32 the angle $\angle HAB = \gamma$ is equal to the angle $\angle ACB$ in the alternate segment in the circle, and similarly, $\angle ABC$

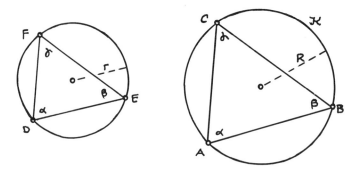

FIGURE 22.1

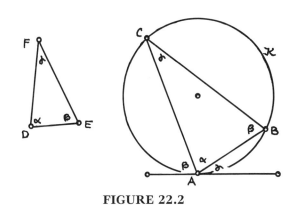

FIGURE 22.2

is equal to β. Because the angles in a triangle add up to two right angles, $\triangle ABC$ is what we want.

Euclid's arguments are most economical; he employs no "higher" tools like similarity arguments. There is only one line of attack, the one using Prop. III. 32. And there is an element of surprise: The angle β is constructed at a location where one really would not look for it. The arguments in this proof are clear and concise; the logic is perfect, and still it is delightfully unexpected. Once more we quote Hardy (§ 10): "... what the public wants is a little intellectual 'kick,' and nothing else has quite the kick of mathematics."

Probably it is this "kick" that G. C. Rota [1997, 121–133] means when he speaks of the "light bulb effect," the sudden understanding of a problem in a flash like a light bulb suddenly being lit. This is not so much what we have seen in Euclid's solution of Prop. IV. 2, but it is a common experience for anybody who ever solved a problem on his own, even when it was as a student in high school. Rota claims that the image of the light bulb is a mistake. He discusses the inter-relations of the three concepts of *truth*, *beauty*, and *enlightenment* in mathematics and holds that the last one is the most important one for a theorem or a proof.

> Verification alone does not give us a clue as to the role of a state-ment within the theory; it does not explain the *relevance* of the statement. In short, the logical truth of a statement does not *en-lighten* us to the *sense* of the statement. *Enlightenment* not truth is what the mathematician seeks when asking, "What is this good for?" ... Every teacher of mathematics knows that students will not learn by merely grasping the formal truth of a statement. Stu-dents must be given some enlightenment as to the sense of the statement, or they will quit. ... Mathematical beauty is the expres-sion mathematicians have invented in order to obliquely admit the phenomenon of enlightenment while avoiding acknowledgment of the fuzziness of this phenomenon. ... The term "mathematical beauty," together with the light bulb mistake, is a trick mathemati-cians have devised to avoid facing up the messy phenomenon of enlightenment. (Rota p. 131/132)

Unfortunately, Rota does not provide us with an analysis of a spe-cific example in order to make his thesis somewhat more concrete. ("Tradition [in phenomenology] demands that no examples ever be given of what one is talking about." (Rota p. 200). This is like Eu-clid in his arithmetical Books.) Some elaborations on mathematical proofs in Rota's subsequent chapter could very well serve his pur-pose, but these are too long and detailed to be quoted here. In any case, enlightenment seems to be just one, albeit important, compo-nent of mathematical beauty. We will try to discuss this component in later examples and will refer back to what was said in this section on suitable occasions.

23

CHAPTER

Euclid Book X

Incommensurable Magnitudes

....

Book X comprises fully one quarter of the *Elements*. Not only is it the most voluminous book, but also the most difficult to read. Setting aside the bulk of the material of Book X, we concentrate our attention on the introductory part of the book, which is of general interest. Some recent studies have done much to elucidate the tedious and long-winded considerations of Book X. The interested reader should look up the papers by Taisbak [1982], Knorr [1985], Fowler [1992], as well as the respective chapters in Mueller [1981] and van der Waerden [1954]. The original motivations for the classification of incommensurable lines apparently can be found in the material presented in Book XIII. During the discussion of the relevant propositions of Book XIII we will return to the problems of Book X.

The discovery of incommensurable lines represents one of the highest achievements of the early Greek mathematicians. The historical development from the time of Pythagoras to Plato will be sketched in the chapter on the origin of incommensurability.

Definitions

Def. X. 1.
Those magnitudes are said to be commensurable which are measured by the same measure, and those incommensurable which cannot have any common measure.

Throughout the *Elements*, Euclid takes measuring to be an undefined notion. The magnitudes of Book X are always lines and rectangles. The concept of incommensurability is refined, for the purposes of Book X, by:

Def. X. 2.
Straight lines are commensurable in square when the squares on them are measured by the same area, and incommensurable in squares when the squares on them cannot possibly have any area as a common measure.

Definitions 3 and 4 allow a glimpse into the latter parts of Book X. From everyday experience one knows that it is very helpful to have a standard measure for measuring lines. This will also help to measure areas, as we have seen in the discussion of Prop. VI. 1. Euclid fixes such a line in Def. X. 3 and calls it the *"expressible* line *a."* (We are adopting the terminology of Fowler [1992]. Heath and others call it the rational line.) Any other line b such that b^2 is commensurable with a^2 will also be called *expressible*. An area that is commensurable with a^2 is called expressible. Lines and areas that are not expressible are called *alogoi*. (According to the Greek *alogos*. Irrational is somewhat misleading.) Examples in modern terminology: Let the standard measure a have length 1. Then a line b of length $\sqrt{5}$ is incommensurable with a, but commensurable in square with a; hence b is expressible. One can construct, starting from a, a line b of length $\sqrt{5}$ and then a rectangle "contained" by a and b of area $\sqrt{5}$. This can be squared by Prop. II.14, and its side of length $\sqrt[4]{5}$ will then be *alogos*. The modern mathematician will immediately recognize that proceeding like this amounts algebraically to successive quadratic extensions of the field of rational numbers. The classification of numbers (or segments) in that way is beyond the means of Greek mathematics.

The Euclidean Algorithm for Magnitudes

Prop. X. 1.
If two unequal magnitudes are set out and if there is subtracted from the greater more than its half and from the remainder more than its half and this goes on forever, there will be left some magnitude which will be less than the lesser magnitude set out. It can be proved similarly even if the things subtracted are halves.

For the modern reader it will be obvious that this is the Archimedean property of the ordering of magnitudes, and Euclid proves the statement by referring to Def. V. 4. In fact, the proof shows that after a finite number of subtractions, the residue will be smaller than the second magnitude. This remark is essential for the next proposition as well as for the limiting procedures in Book XII.

Prop. X. 2.
If, when two unequal magnitudes are set out and the lesser is always subtracted in turn from the greater, the remainder never measures the magnitude before it, then the magnitudes will be incommensurable.

In other words: If the Euclidean algorithm is performed with two magnitudes and does not terminate after finitely many steps, then the two magnitudes will be incommensurable. It is usually called, by its Greek name, the *anthyphairesis* procedure. Evidently, this is a useful criterion for incommensurability, but it is used neither in Euclid's *Elements* nor elsewhere in Greek mathematics. One need only know that the algorithm becomes eventually periodic. We will show how easily this can be seen for some familiar lines from elementary geometry (cf. next section).

The next two propositions resemble even more closely their partners, Props. VII.2/3:

Props. X. 3/4.
Given two (three) commensurable magnitudes to find their greatest common measure.

Proof:
If the two magnitudes a, b are commensurable, the Euclidean algorithm performed with them will terminate (Prop. X. 2). Let it terminate with a last remainder f measuring the one before it. Then f will be the greatest common measure of a and b.

The first four propositions have no further applications in Book X. They may be a relic from an earlier theory.

Magnitudes and Numbers

Let again a and b be two commensurable magnitudes. From the definition of commensurability we have a common measure c of a and b and numbers k and m such that $a = kc$ and $b = mc$. This gives us

$$a : b = kc : mc.$$

The modern reader would continue with

$$kc : mc = k : m,$$

and so does Euclid. He ignores that he has two different definitions of proportionality for magnitudes, Defs. V. 5/6, and for numbers, Def. VII. 20. Like Euclid we will skip over these difficulties and just quote

Props. X. 5/6.

$$\left. \begin{array}{c} \textit{The magnitudes } a, b \\ \textit{are commensurable} \end{array} \right\} \Leftrightarrow \left\{ \begin{array}{c} \textit{There exist numbers } k, m \textit{ such that} \\ a : b = k : m. \end{array} \right.$$

Disregarding the conceptual problems, we recognize that X. 5/6 (and their contrapositives 7/8) are a most useful criterion for commensurability. This criterion is expanded for the further theory in Book X in

Prop. X. 9.

$$a : b = k : m \qquad \Leftrightarrow \qquad a^2 : b^2 = k^2 : m^2.$$

23.1 Commensurability and Its Relation to Other Notions

In propositions X. 10–20 Euclid systematically explores the properties of commensurability and its interaction with earlier concepts. The examples constructed in Prop. X, 10 show that his theory is not about the empty set. Next, commensurability is shown to be consistent with the theory of proportions: If a and b are commensurable and $a : b = c : d$, then c, and d will also be commensurable (X, 11). Commensurability is transitive and hence, as we might add, an equivalence relation (X. 12/13). The interaction of commensurability with the addition and subtraction of lines is the theme of X. 14–16. The geometric equivalents of multiplication and division are treated in X. 19/20.

How does commensurability behave with respect to the application of areas? One could translate this into the following problem: When does a quadratic equation with rational coefficients have rational solutions? The answer to this question is presented in Props. X. 17/18. The whole setup of these propositions is similar to Diophantus's Problem 27 (cf. the section on Diophantus). This is apparent from the lemma preceding Prop. 17, where we see a rectangle with sides x, y, area $xy = A$, and half-perimeter $x + y = B$. We abbreviate Euclid's statement a little and introduce our notation along with the text of the proposition (a COMS b means that a and b are commensurable).

Prop. X. 17.
If there are two unequal straight lines $b > c$ and to b is applied the area $\frac{c^2}{4}$ (i.e., with defect) and deficient by a square (that means $bx - x^2 = \frac{c^2}{4}$),

*and if $r^2 = c^2 - b^2$ (to be constructed via Pythagoras's theorem), then
we will have*

$$x \text{ COMS } b \qquad \Leftrightarrow \qquad r \text{ COMS } b.$$

Clearly, Euclid's proof is geometric and uses Prop. II.5 as a substitute for our binomial formula, as well as many earlier propositions about proportions and the properties of commensurability. Translated into algebraic shorthand it becomes transparent for the modern reader. We have

$$bx - x^2 = \frac{c^2}{4},$$
$$x^2 - bx + \frac{b^2}{4} = \frac{b^2}{4} - \frac{c^2}{4},$$
$$\left(x - \frac{b}{2}\right)^2 = \frac{1}{4}(b^2 - c^2),$$
$$x - \frac{b}{2} = \frac{1}{2}\sqrt{b^2 - c^2} = \frac{1}{2}r.$$

From this, the assertion is obvious. We see that Euclid has the same clear understanding of quadratic problems long before Diophantus.

With this we take our leave of Book X. However, there will be a few remarks in the discussion of Book XIII, as results from Book X are used there.

24

CHAPTER

The Origin of Mathematics
12

Incommensurability
and Irrationality

....

Aristotle, near the beginning of his *Metaphysics*, a most prominent place, quotes incommensurability as the prototype of a scientific discovery. After a short general description of philosophy he writes:

> [The first philosophers] were pursuing science in order to know, and not for any utilitarian end.... Evidently then we do not seek it [philosophy] for the sake of any other advantage; but as the man is free, we say, who exists for himself and not for another, so we pursue this as the only free science, for it alone exists for itself [982b 20–27].... All the sciences, indeed, are more necessary than this [philosophy], but none is better. Yet the acquisition of it must in a sense end in something which is the opposite of our original inquiries. For all men begin, as we said, by wondering that the matter is so (as in the case of ... the solstices or the incommensurability of the diagonal, for it seems wonderful to all men who have not yet perceived the explanation that there is a

thing which cannot be measured even by the smallest unit). But we must end in the contrary and, according to the proverb, the better state, as is the case in these instances when men learn the cause; for there is nothing which would surprise a geometer so much as if the diagonal turned out to be commensurable. [Metaphysics 983a 10–21]

Men are able to advance knowledge by pure reasoning alone. That is why incommensurability and mathematics as a whole are of such paramount importance for Plato, Aristotle, and so many other philosophers.

Many papers and books have been written about the history of incommensurability, for instance Knorr [1975] and Fowler [1987]. We will confine ourselves to the most basic historical dates and to the more mathematical aspects of the story.

Information about the early history of incommensurability comes from relatively late sources. Pappus (ca. 330 c.e.) wrote a commentary on Euclid's Book X from which the following fragment is quoted by Heath in his introduction to Book X (Euclid–Heath vol. 2 p. 3): The theory of irrational magnitudes "had its origin in the school of Pythagoras. It was considerably developed by Theaetetus, the Athenian . . .". Concerning the Pythagoreans we have some more vague reports from about 700 or 800 years after the original discovery was made. About 290 c.e. Iamblichus tells some stories about the Pythagorean Hippasus of Metapontum (ca. 500–450 b.c.e., quoted after Knorr [1975] p. 50 n. 6/7). Hippasus concerned himself with the "sphere of the twelve pentagons," that is, the dodecahedron. He made the construction public and took credit for this himself (instead of giving it to Pythagoras). As a punishment for this he perished at sea. Another version of this account metes out this same punishment to the one who divulged knowledge of the irrational. Commentators have called this tradition confused and unreliable, but I concur with the opinion expressed by v. Fritz [1945] and believe it has its origin in actual mathematical discoveries. The construction of the dodecahedron is based on the ratio $d : f$ of the diagonal and side of the regular pentagon, yet d and f are incommensurable, as will be shown below. For mathematical reasons the stories about Hippasus seem plausible in spite of the scanty historical sources.

In the following subsections the main stages of the development of the theory of irrationals are described in some detail. At the earliest Pythagorean stage (about 500–450 B.C.E.) we will talk about the side and the diagonal of the square and the regular pentagon. For stage two we consider the theorem of Theodorus (about 430 B.C.E.) concerning the irrationality of $\sqrt{3}, \ldots, \sqrt{17}$. In stage three Theaetetus, a student of Theodorus, created the foundations of the theory preserved in Euclid's Book X.

Arithmetical Proofs: The Square

Recall the definition of commensurability and two criteria equivalent to the definition: to have a terminating Euclidean algorithm, or *anthyphairesis*, procedure of two magnitudes a, b, as defined in Prop. X. 2, or to have a ratio $a : b = k : m$ as "number to number." We will present two proofs in this section using the latter criterion. The first one comes from an appendix to Euclid's Book X, which is not included in Heath's translation and is usually called Prop. X. 117. It has nothing to do with the other material of Book X. Moreover, it has a short introductory note at the beginning of the proof explaining the overall strategy, something never found in Euclid's other proofs. This note is in close agreement with a remark by Aristotle:

> For all who effect an argument *per impossible* deduce what is false, and prove the original conclusion hypothetically when something impossible results from the assumption of its contradictory, e.g. that the diagonal of the square is incommensurate with the side, because odd numbers are equal to evens if it is supposed to be commensurate. (*Prior Analytics* 41a 23–27)

We may therefore assume that Prop. X. 117 is a later addition to Euclid's text that was important for its relation to Aristotle's teachings and, even more so, for historical reasons. The proof itself looks very archaic and may well, in its essence, go back to the Pythagoreans. We present the proof in the translation of v. Fritz [1945, p. 254 n. 60] together with our comments.

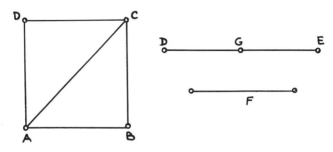

FIGURE 24.1

Prop. X. 117 (= Appendix 27).

Let it be assigned to us to prove that in the case of square figures the diameter is incommensurable in length with the side.

Let *ABCD* be a square and *AC* its diameter. I say that *AC* will be incommensurable with *AB* in length (Fig. 24.1).

In spite of the artfully interwoven contradictions, there is essentially only one line of attack, as Hardy says, and the proof carries a huge potential for generalization. One just has to replace, for instance, the number 2 by any prime number and the minimi by unique prime factorization.

The Proof	Comments
(i) For let us assume that it is commensurable. I say that it will follow that the same number is at the same time even and odd.	(i) In contrast to Euclid's habits, the outline of the proof is sketched.
(ii) It is clear that the square on AC is double the square on AB.	(ii) Three basic facts: (a) The special case of the theorem of Pythagoras I. 47 is used.
Because then (according to our assumption) AC is commensurable with AB, AC will be to AB in the ratio of an integer to an integer.	(b) The criterion Prop. X. 5 is quoted.
Let them have the ratio DE : F and let DE and F be the smallest numbers which are in this proportion to one another.	(c) The minimi of Prop. VII. 20 are going to be used. One number is designated by DE because below it will be bisected at G. This is Euclid's usual notation.
(iii) DE cannot then be the unit. For if DE were the unit and were to F in the same proportion as AC to AB, AC, being greater than AB, DE, the unit, will be greater than the integer F, which is impossible. Hence DE is not the unit, but an integer (greater than the unit).	(iii) Because the unit is not a number according to Euclid's definition, the case DE = 1 has to be dealt with separately. Moreover, he needs this because later he has to consider half of DE.
	First use of the method of contradiction.
	(This is in fact a fake contradiction like so many others found in textbooks. Observe that he has called DE a number from the outset. The terminology concerning the unit is not consistent.)

(iv) Now since $AC : AB = DE : F$, it follows that also $AC^2 : AB^2 = DE^2 : F^2$. But $AC^2 = 2AB^2$ and hence $DE^2 = 2F^2$. Hence DE^2 is an even number ...

(iv) The theorem of Pythagoras and some decisive properties of proportions (VI. 20 and VIII. 11) are used to obtain the key element of the proof.

(v) ... and therefore DE must also be an even number. For if it were an odd number its square would also be an odd number. For if an odd number of odd numbers is added, the whole is odd.

(v) This is a repetition of Prop. IX, 23

Second contradiction

(vi) Hence DE will be an even number. Let then DE be divided into two equal numbers at the point G. Since DE and F are the smallest numbers which are in the same ratio they will be relatively prime. Therefore, since DE is an even number, F will be an odd number. For if it were even, the number 2 would measure both DE and F, though they are relatively prime, which is impossible.

After Prop. VII 22

Third contradiction

(vii) Hence F is not even, but odd. Now since $ED = 2EG$ it follows that $ED^2 = 4EG^2$. But $DE^2 = 2F^2$, and hence $F^2 = 2EG^2$. Therefore F^2 must be an even number, and in consequence F also an even number. But it has also been demonstrated that F must be an odd number, which is impossible.

The argument from part (iv) has to be repeated.

Fourth contradiction

(continued)

(continued)

(viii) It follows, therefore, that
AC cannot be commensurable
with AB, which was to be
demonstrated.

Philosophical comment by Plato (*Republic* V, 454 a 1):

"What a great thing, Glaucon, is the power of the art of contradiction."

Poetic comment by Rainer Maria Rilke:

Rose, o pure contradiction, delight,
To be nobody's sleep under so many eyelids.

(Rose, oh reiner Widerspruch, Lust,
Niemandes Schlaf zu sein unter so viel Lidern.)
Inscription on Rilke's tombstone.

Arithmetical Proofs: The Pentagon

Let d be the diagonal and f the side of a regular pentagon. From the second analysis of the pentagon in our discussion of Book IV (Fig. 11.6 and Prop. XIII. 8) we recall the basic proportion

$$d : f = f : (d - f).$$

Now let us assume that d and f are commensurable and k and m are minimi such that

$$d : f = k : m.$$

Then we have immediately

$$k : m = d : f = f : (d - f) = m : (k - m),$$

where m and $k - m$ are smaller than k and m, respectively; a contradiction. Hence d and f are incommensurable. As soon as one occupies oneself a little with the pentagon and has the notion of a ratio in least terms, this proof is very obvious and is simpler than the one for the square. It uses only the fundamental property of minimi and needs no further elaboration. It is, however, not to be found in any source from antiquity.

Geometrical Proofs: The Square

Euclid never uses his first criterion of the nonterminating anthyphairesis procedure (Prop. X. 2) for proofs of incommensurability. We are going to do just this, first for the square and then for the pentagon. Even if Euclid does not apply this criterion in any particular instance, we do have some evidence that it was in fact used in times before Euclid. The first point is that Euclid has preserved it. The second is a much-quoted passage from Aristotle that shows that there was an anthyphairesis–related definition of ratio in use before Eudoxus created the newer theory of Book V. Aristotle speaks about a pre–Euclidean proof of Prop. VI. 1. (he uses the term antanairesis, which is equivalent to anthyphairesis): "But once the definition is stated, the said becomes immediately clear. For the areas and the bases have the same antaneiresis; such is the definition of the same ratio." According to Aristotle, then, the anthyphairesis procedure was even more generally used than just for proofs of incommensurability. We will not enter into this general field and merely present the common reconstructions for some simple pairs of lines.

Consider again the side $AB = a$ and the diagonal $AC = d$ of a square (Fig. 24.2).

The circle centered at C with radius $CB = AB$ cuts the diagonal AC at G. Let GF be the tangent of this circle. Using the notation $CA = d$, $AB = a$, $AG = a_1$, we see that $a > \frac{1}{2}d$ and $d = a + a_1$ and $a_1 < a$.

In the next step we have to subtract AG from $a = AB$. We claim that $AG = FB$. To see this, consider the triangles $\triangle FBC$ and $\triangle FGC$. They are congruent because they have the common side FC, $BC =$

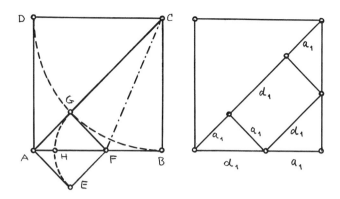

FIGURE 24.2

GC, and the right angles at B and G. Hence $GF = BF$. But obviously, $GF = AG$. Now subtract $AG = GF$ from FA as indicated, with the result

$$a = AB = 2AG + \text{ remainder } AH.$$

The process is now repeated using the square $AGFE$ with H in the role of G, so that we find exactly the same situation that we had before in $ABCD$. Consequently, the procedure becomes periodic and will never terminate. AC and AB are incommensurable.

Even if we do not have any direct witness for the above proof, there is some indirect evidence relating to this or an equivalent procedure. In order to understand this we need a little preparation. Instead of iterating to smaller and smaller squares, we will go in the other direction, from smaller to bigger squares. Call the diagonal AF of the little square d_1 and its side a_1. We have shown that

$$a = d_1 + a_1,$$
$$d = d_1 + 2a_1.$$

Now let $a = a_2$ and $d = d_2$, and iterate to a larger square,

$$a_3 = d_2 + a_2,$$
$$d_3 = d_2 + 2a_2,$$

and so on.

Now let us turn to numbers—that is, integers—and define a sequence a_i, d_i by

$$a_{i+1} = d_i + a_i,$$

$$d_{i+1} = d_i + 2a_i.$$

Starting with $a_0 = 1$ and $d_0 = 1$, we obtain

i	0	1	2	3	4	5	$\ldots$
a_i	1	2	5	12	29	70	
d_i	1	3	7	17	41	99	
$2a_i^2$	2	8	50	288	1682	9800	
d_i^2	1	9	49	289	1681	9801	

This is what the Greeks have called the side and diagonal numbers. Plato, in a rather obscure passage in the *Republic* 546 c, speaks about something like 7 being "the rational diameter" of 5, and his commentators from antiquity ascribe this sequence to the Pythagoreans. By induction one may prove that $2a_i^2$ always misses d_i^2 by 1 or -1. In any case, this sequence was well known, and it is closely related to the anthyphairesis procedure, however in the direction going to greater instead of smaller segments. (See the Notes for further details.)

Geometrical Proofs: The Pentagon

The geometrical proof for the pentagon is much easier than the one for the square. Let $d = AC$ be the diagonal and $f = AB$ the side of the regular pentagon $ABCDE$ (Fig. 24.3).

In the figure we recognize $f = AW$ and $d - f = WC = ZX = d_1$, the diagonal of the smaller pentagon $VWXYZ$, and its side $VW = VC - WC = f_1$. The anthyphairesis procedure is

$$d = f + d_1,$$
$$f = d_1 + f_1.$$

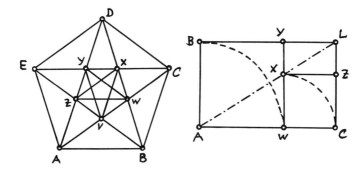

FIGURE 24.3

From similar figures we have $d_1 : f_1 = d : f$, and the procedure becomes periodic, producing a sequence of smaller and smaller nested pentagons. The similarity of the pentagons is reflected in the similarity of the rectangles $\square ACLB$ and $\square XZLY$. The latter similarity is by Props. VI.24/26, equivalent to the fact that the point X is on the diagonal AL of the greater rectangle.

This procedure is closely related to Prop. XIII. 5. Even if we have no written source for the resulting sequence, evidently it was used in a famous Greek building. As before, we construct side and diagonal numbers, this time for the pentagon. Starting from $f_0 = 1 = d_0$ we derive the sequence

i	0	1	2	3	4	5	...
f_i	1	2	5	13	34	89	
d_i	1	3	8	21	55	144	

The sequence 1, 1, 2, 3, 5, 8, ... is the well-known Fibonacci sequence. Archeologists have pointed out that the numbers 13, 21, 34, 55 are realized in the plan of the ancient Greek theater in Epidaurus. Epidaurus was a kind of health-care center. The ground plan of its theater is based on a hidden pentagonal symmetry. (The half-circle is divided into 10 equal parts by the stairs.) This relates well to the pentagon as a medical symbol or the Pythagorean pentagram, which was called "health." We show the ground plan (Fig. 24.4) according to Käppel [1989], who quotes it from v. Gerkan–Müller Wiener [1961].

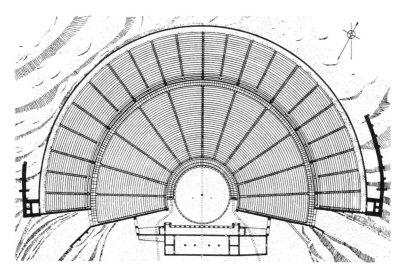

FIGURE 24.4

The lower part of the theater has 34 rows of seats and 13 stairs. The upper part has 21 rows of seats. The actual construction of the theater took place at two separate periods. The lower part was built about 300 B.C.E., that is, at the time of Euclid, and the upper part was completed about 170 B.C.E. Käppel demonstrates convincingly that the completion of the theater followed the original plan from 300 B.C.E. It would be very hard to believe that the occurrence of Fibonacci numbers in a building with a fivefold symmetry would be by pure chance. Even if we don't have any written sources, it seems clear that the Fibonacci numbers have been derived somehow from the anthyphairesis procedure of the pentagon.

The Mathematician Theodorus of Cyrene

According to Eudemus, Theodorus of Cyrene (in Libya) was a contemporary of Hippocrates of Chios. In his *Memorabilia* (of Socrates VI.2) Xenophon, who was a student of Socrates, says that Theodorus was a famous geometer, and Diogenes Laertius (about 300 C.E.) calls him the teacher of Plato. In Plato's own dialogues such an amiable picture of Theodorus is drawn that Diogenes Laertius's claim becomes very convincing. Theodorus and Socrates appear to be of about the same age in the dialogue "Theaetetus," which plays in the year 399 B.C.E. shortly before Socrates' death. From all this we may conclude that Theodorus lived from about 465 B.C.E. until some time after 400 B.C.E. At least for a time he resided in the intellectual center Athens, where Socrates' students Xenophon and Plato met him. If he was born around 465, then he may well have made his major mathematical discoveries in the years 440–420, which is in good agreement with what Eudemus says.

Theodorus started out as a student of philosophy with Protagoras (ca.485–415), but as Plato reports him saying:

> My own inclinations diverted me at rather an early age from abstract discussions to geometry. (Theaetetus 165a)

Plato has Socrates asking the student Theaetetus about Theodorus:

> S.: Is Theodorus a painter?
> Th.: Not so far as I know.
> S.: Nor an expert in geometry either?
> Th.: Of course he is, Socrates, very much so.
> S.: And also in astronomy and calculation and music and in all the liberal arts?
> Th.: I am sure he is.
> (Theaetetus 145a)

This is the historically first instance where the classical liberal arts arithmetic, geometry, astronomy, and music are mentioned, and moreover Theodorus as an expert in all of them.

At the beginning of the dialogue, the student Theaetetus states one of Theodorus's great achievements:

The Theorem of Theodorus

> Theaetetus: *Theodorus here was drawing diagrams to show us something about powers—namely that a square of three square feet and one of five square feet aren't commensurable, in respect of length of side, with a square of one square foot; and so on, selecting each case*

individually, up to seventeen square feet. At that point he somehow got tied up. (Theaetetus 147d)

The statement of the theorem is very clear. The segments of length $\sqrt{3}$, $\sqrt{5}$, ..., $\sqrt{17}$ are incommensurable with a segment of unit length, or in other words, are irrational (with the obvious exceptions). Like every commentator of this passage we may say that Plato did not include $\sqrt{2}$ in his list because the respective proof was known before Theodorus's theorem. The second question is, What does it mean that he somehow got tied up at the number 17? and the third one, What kind of proof was Theodorus able to give for his statement? Many kinds of reconstructions of the proof have been proposed by various authors. The ones up to 1974 are listed and discussed by Knorr [1975], who adds his own attempt at an arithmetical proof. An overview for the time 1975–1993 can be found in my paper Artmann [1994]. Naturally, I am convinced that my own hypothesis about the historical proof is the most plausible one. I will sketch it and refer the interested reader to my paper for more details.

The reconstruction will use a suitable geometric version of the criterion X. 2. In modern terms this will be equivalent to the study of the continued-fraction expansion of the roots, but, save for a few notational abbreviations, no use will be made of modern methods. H. G. Zeuthen [1915] was the first to propose the use of continued fractions as an explanation for Theodorus's success up to $\sqrt{17}$. Basically, then, the idea of this geometric proof is due to Zeuthen. It is modified in such a way that it uses nothing but the simplest parts of similarity geometry for rectangles, which are all contained in Euclid's *Elements*, Book VI.

We explain the geometric version of the procedure, or rather the algorithm of anthyphairesis, anthyphairesis, for short, in the simple case of a rectangle with sides $a = 16$ (units) and $b = 7$ (units).

Look at the rectangle contained by a and b with $a > b$. Take b away from a as often as possible to get the representation with $r_1 < b$ and an integer $q_0 \geq 1$,

$$a = q_0 b + r_1,$$
$$\text{here} \quad a = 2b + 2.$$

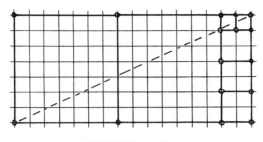

FIGURE 24.5

Now look at the smaller rectangle contained by b and r_1 to obtain

$$b = q_1 r_1 + r_2,$$
$$\text{here} \quad 7 = 3r_1 + 1.$$

Proceed with

$$r_1 \ = \ q_2 r_2 + r_3,$$
$$\text{here} \quad 2 \ = \ 2r_2 + 0,$$

and so on.

If we start, as in this case, with commensurable magnitudes a, b and numbers k, m such that

$$a : b = k : m,$$

then anthyphairesis is nothing but the Euclidean algorithm in a geometric version and will end after finitely many steps. Observe that each time we take away more than half of the respective magnitude. Starting the same algorithm with two arbitrary segments $a > b$ will give us a sequence of remainders r_i and quotients q_i. If some r_i turns out to be 0, then we are in the commensurable case: otherwise not. From hindsight and in modern terms, the anthyphairesis coefficients $[q_0; q_1, q_2, \ldots]$ of $a : b$ are the coefficients of the continued fraction expansion for $a : b$. Lagrange has shown that if $a = rb$ (with some real number r), then the sequence $[q_0; q_1, q_2, \ldots]$ will be (infinite and) periodic if and only if r is the irrational root of a quadratic equation with integer coefficients [Hardy–Wright 1954, § 10.12].

Thus our geometric method can be (theoretically) successful for numbers $r = \sqrt{n}$. This is sufficient for Theodorus's theorem. It

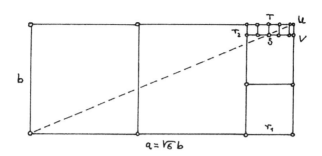

FIGURE 24.6

makes no sense, however, for numbers like $\sqrt{\sqrt{7} - \sqrt{5}}$, which are easily constructed by Euclidean means and are studied in Book X.

We will show how the method works in two particular cases.

Case 1. Start with a segment b and construct with the aid of Pythagoras's theorem a segment a such that $a^2 = 6b^2$. (Use $\sqrt{2}b$ as an intermediate step.) Draw the rectangle contained by a and b and its diagonal f (Fig. 24.6).

We get

$$
\begin{aligned}
a &= 2b + r_1, \\
b &= 2r_1 + r_2, \\
r_1 &= 4r_2 + r_3.
\end{aligned}
$$

Drawing the segments corresponding to the last equation, we observe that the vertex S of the small rectangle $STUV$ with sides $(r_1 - 2r_2)$ and r_2 seems to be on the diagonal f of the original big rectangle. If this were true, then the small rectangle would be similar to and have the same anthyphairesis as the big one; and inside the small rectangle should again appear an even smaller similar one . . . and so on. We would have a periodic anthyphairesis of the form

$$
\begin{aligned}
a &= 2b + r_1, \\
b &= 2r_1 + r_2, \\
r_1 &= 4r_2 + r_3, \\
&\cdots
\end{aligned}
$$

$$r_k = 2r_{k+1} + r_{k+2},$$
$$r_{k+1} = 4r_{k+2} + r_{k+3},$$
$$\ldots$$

with a sequence of coefficients

$$a : b = [2; 2, 4, 2, 4, 2, 4, \ldots, 2, 4, \ldots].$$

In particular, the anthyphairesis for $a = \sqrt{6}b$ and b would not terminate, and the segments a and b would be incommensurable.

From Fig. 24.6 we can see that

(*) $$r_1 = a - 2b \quad \text{and} \quad 0 < r_1 < b,$$

(**) $$r_2 = b - 2r_1 = 5b - 2a \quad \text{and} \quad 0 < r_2 < r_1.$$

These geometrical relations imply that

$$r_1 - 2r_2 = 5a - 12b.$$

Let us now check whether what the drawing suggests is true. "S on the diagonal" is equivalent to the similarity of the two rectangles (Props. VI. 24/26), hence to

(1) $$(r_1 - 2r_2) : r_2 = a : b,$$

and this in turn is equivalent (Props. VI. 16) to

(2) $$(r_1 - 2r_2)b = ar_2.$$

"Analysis" proceeding from (2) with (*), (**) gives

$$(5a - 12b)b = a(5b - 2a),$$
$$5ab - 12b^2 = 5ab - 2a^2,$$
$$6b^2 = a^2$$

We have found our starting point. Going backward with "synthesis" proves "S on the diagonal" and hence the incommensurability of b and $a = \sqrt{6}b$.

Let us sum up what we have done. Guided by the drawing, we guessed equation (1), which would give us incommensurability. The guess was verified by reducing (1) to the equation $a^2 = 6b^2$, which was what we started with. Even if the verification (synthesis) constitutes the exact proof, we would have been completely at a loss without the guess from the drawing.

Except for the verifications of $2b < a < 3b$ and similar relations that could easily be supplied by starting from $a^2 = 6b^2$, we have a perfectly valid proof for the incommensurability of a and b, or, in other words, for the irrationality of $\sqrt{6}$.

This same method works equally well for all the roots $\sqrt{3}, \ldots,$ $\sqrt{17}$, with the exception of $\sqrt{13}$, which will be our case 2. After case 2 we will say a few words about precise drawings and related things, but first we have to discuss $\sqrt{19}$.

From the modern theory of continued fractions we know the sequence of anthyphairesis coefficients for $\sqrt{19}$ to be $[4; \overline{2, 1, 3, 1, 2, 8},$ $\ldots]$ with a period of length 6. Suppose we started with a unit length of 1m (and therefore $\sqrt{19} \approx 4.3589$m). The size of the small rectangle similar to the big one would be (after 6 steps in the procedure) about 1mm x 4mm. It is thus impossible to guess the period from an actual drawing. The method breaks down after $\sqrt{17}$. Observe that $\sqrt{18} = 3\sqrt{2}$ plays no significant role. This explains why Theodorus got tied up after $\sqrt{17}$.

Case 2. The true sequence of coefficients for $\sqrt{13}$ presents the same problems as the one for $\sqrt{19}$. It is $[3; \overline{1, 1, 1, 1, 6}, \ldots]$ with a period of length 5. Even worse, the odd length of the period would give us a point on the diagonal only after two cycles, which is impossible to find graphically. In spite of these problems there is an easy solution if we look at it the other way round. There are some especially simply sequences like

$$(1 + \sqrt{5}) : 2 = [1; 1, 1, 1, \ldots],$$
$$\sqrt{2} : 1 = [1; 2, 2, 2, \ldots],$$
$$\sqrt{5} : 1 = [2; 4, 4, 4, \ldots],$$
$$\sqrt{10} : 1 = [3; 6, 6, 6, \ldots],$$

but a sequence like $[n; 3, 3, 3, \ldots]$ is missing. Can we find a and b such that the rectangle contained by a and b would produce the sequence $[3; 3, 3, 3, \ldots]$? A rectangle with this anthyphairesis would look like the one in Figure 24.7, with the point S on the diagonal and the first remainder x unknown. For the sake of simplicity, let $b = 1$. By

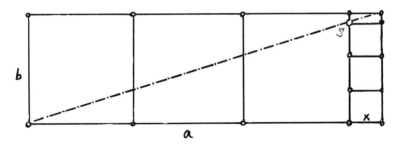

FIGURE 24.7

Props. VI. 24, 26, 16 we have

$$\begin{aligned}
S \text{ on the diagonal } &\Leftrightarrow & 3 : 3x &= x : (1 - 3x) \\
&\Leftrightarrow & 3(1 - 3x) &= 3x^2 \\
&\Leftrightarrow & (3 + x)x &= 1
\end{aligned}$$

To solve this for x is a standard problem of the application of areas with excess (Prop. VI. 29 in its simplified version), see Fig. 24.8.
The result is

$$x + \frac{3}{2} = \frac{\sqrt{13}}{2}.$$

Hence

$$a = 3 + x = \frac{3}{2} + \frac{\sqrt{13}}{2}.$$

This shows that a is incommensurable to $b = 1$, and hence $\sqrt{13}$ is incommensurable to 1 by Props. X. 15/16. (Clearly, Euclid is a

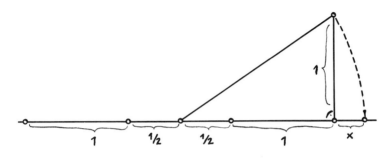

FIGURE 24.8

later source, but the statement "x is commensurable to b if and only if $3b + x$ is commensurable to b (or $2b$)" is trivial from the definition of commensurability.)

It was more or less by chance that $\sqrt{13}$ turned up by starting from this sequence, but anyway it completes the geometrical proof of Theodorus's theorem. On the other hand, one can turn the trick into a method (a general feature of mathematics) and derive an infinite sequence of segments incommensurable with the unit segment by starting from sequences like $[k; k, k, k, \ldots]$ with some number k.

A few words have to be said about the practice of drawings. For teaching purposes it seems that drawings in the smoothed-over sand on the floor of the gymnasium played the role of our chalk and blackboard. Such drawings are not accurate enough for the proof presented above. Even if no expert draftsmanship is necessary for the practical execution of the anthyphairesis for the respective rect-angles, one has to be careful. It was only after some futile attempts I became skillful enough for precise drawings. After that, the method works surprisingly well. I started with segments b of about 10 cm length and constructed a with the help of Pythagoras's theorem, e.g., $15 = 4^2 - 1^2$ or $13 = 3^2 + 2^2$. The precise length of a turns out to be the most important parameter in the procedure. When actually doing the drawings, I was quite surprised by the simplicity of the case $\sqrt{7}$ and the difficulties for $\sqrt{10}$. This is something one cannot learn from theory.

Returning to antiquity, we know that very accurate drawings were done for architectural purposes. Heisel [1993] presents a gen-eral overview. Petronotis [1969, p. 4] observes the introduction of a new, very precise method of on–site drawings of the ground plan of temples incised in the floor before the middle of the sixth cen-tury B.C.E. The Hellenistic construction plans for the columns and other parts of the temple of Apollo at Didyma, incised in the walls, have been published by Haselberger [1985]. Plato hints at precise drawings in his *Philebus* 56b, where he praises the exactness of archi-tecture and shipbuilding. In late antiquity, St. Augustine remarks, in a passage where he represents neoplatonic views of arithmetic and geometry (*Confessiones* Book X. 12): "I have seen the lines of archi-tects, the very finest like a spider's thread. . . ." Hence it would pose no problem for Theodorus to do (or have done) drawings with the desired accuracy on wood tablets or precisely planed marble slabs as used in the building trade. A unit length of one foot (about 30cm) would have been entirely feasible.

This completes our discussion of the theorem of Theodorus.

Theaetetus of Athens

Theaetetus of Athens studied under Theodorus and at the Academy with Plato. Plato made him a principal character in his dialogue "Theaetetus," and what we know about the life of Theaetetus is chiefly derived from the dialogue. Since he is characterized as a young student who understands more about mathematics than a great many men with long beards (168e), we may conclude that he was about 15 or 16 years old at the time of the dialogue, that is, 399. The introduction to the same dialogue reports him dying in 369 B.C.E. Plato has Theodorus describe his student to Socrates:

> I assure you that, among all the young men I have met with—and I have had to do with a good many—I have never found such admirable gifts. The combination of a rare quickness of intelligence with exceptional gentleness and of incomparable virile spirit with both, is a thing that I should hardly have believed could exist, and I have never seen it before ... his approach to learning and inquiry, with the perfect quietness of its smooth and sure progress, is like the noiseless flow of a stream of oil. It is wonderful how he achieves all this at his age. (*Theaetetus* 144 a, b)

In the dialogue Socrates, Theodorus, and Theaetetus discuss what knowledge is. Two mathematical statements are used as examples of knowledge: Theodorus's theorem, which we have investigated, and immediately following a definition and statement found by Theaetetus together with his friend, the younger Socrates. Remember that Theodorus stopped for some reason at $\sqrt{17}$. Theaetetus continues:

> The idea occurred to us, seeing that these square roots were evidently infinite in number, to try to arrive at a single collective term

by which we could designate all these roots.... We divided number in general into two classes. Any number which is the product
of a number multiplied by itself we likened to the square figure,
and we called such a number "square" or "equilateral." ... Any
intermediate number, such as three or five or any number that
cannot be obtained by multiplying a number by itself, but has one
factor either greater or less than the other, so that the sides containing the corresponding figure are always unequal, we likened
to the oblong figure, and we called it an oblong number.... All the
lines which form the four equal sides of the plane figure representing the equilateral number we defined as *length*, while those
which form the sides of squares equal in area to the oblongs we
called *roots* [surds], as not being commensurable with the others
in length, but only in the plane areas to which their squares are
equal. And there is another distinction of the same sort in the case
of solids. (*Theaetetus* 147d–148b)

The two students have gone three steps beyond the theorem of
Theodorus: (i) They gave the geometric definition of square and
oblong numbers; (ii) they seem to have proved the irrationality of
$\sqrt{n}$ for any nonsquare number n; (iii) they introduced the notion of
lines "commensurable in square", that is, Def. X. 2. For the proof of
(ii) we have no direct evidence, it may have been the arithmetical
criterion for incommensurability. Point (iii) is a clear confirmation
of the scholion that credits Theaetetus with the original beginnings
of Book X. We have already said that we will return to these things
during the discussion of the regular solids, the other great mathematical achievement of Theaetetus. Some possible intermediate stages
between the original draft by Theaetetus and the final version of
Euclid's Book X are depicted by Knorr [1985, pp. 32–35].

We may add a few words about the connections of the above
mathematical examples to philosophy. The central point is that men
are able to gain knowledge by the power of pure reasoning alone.
This is what makes incommensurability so important for Aristotle,
as we have seen above.

Burnyeat [1978, pp. 509–512] discusses in detail the philosophical
meaning of the above mathematical passage in Plato's philoso-

phy. In particular, he points out the decisive role of definitions in mathematics as a model for philosophical investigations.

The prototypical role of geometry for Plato's investigations in the *Theaetetus* on the nature of knowledge has been studied thoroughly by Friedländer [1960, pp. 131–172, esp. 135]. This author, however, avoids specific mathematical details (p. 139, note 19). For him, mathematics contributes from the outset of the dialogue the most important fact that knowledge is indeed possible (p. 135), even if we do not know a definition of knowledge.

There may be still another bridge from the mathematical example to the main theme of the dialogue, admittedly somewhat superficial, but nevertheless to the point. Plato devotes a substantial part of the *Theaetetus* to the discussion of Protagoras's claim that "Man is the measure (metron) of all things" (The homo mensura thesis, *Theaetetus* 152a etc.). Taken literally, this would mean that there is a common measure (metron) for all things. But Theodorus has shown that there are segments with no common measure (metron). Did he not, in that subtle way, disprove the thesis of his teacher Protagoras'? Certainly there is a much deeper sense to the problem, even for geometry: Theodorus himself is declared to be the "metron" of geometric diagrams, i.e., the judge of their validity (169a). But even in its simple geometric sense, the incommensurability proof is a nice example for Friedländer's observation on p. 146: "The geometer refutes the sophist, but he is unaware of doing so." Because Friedländer did not look at the mathematical contents, he did not see, in this particular example, how the geometer disproved the sophist.

Summary

Let us sum up in a global picture what we have learned about the Greek experience with incommensurable segments. The following path of development seems to be plausible, but we should be cautious. Everyone familiar with elementary geometry knows about the many different ways of arriving at a particular result. Even if Theodorus used anthyphairesis, his figures may have looked very different from the ones above. What is true for geometry applies equally well to history. The real historical landscape is even messier than the geometrical one. We have no sources that would tell us about details, false starts, or discarded theories of competing mathematicians or schools.

Stage 1

Discovery of the phenomenon of incommensurable segments in one or both of two special cases: the diagonals and sides of the square and the regular pentagon. I see no way to decide which one came first and whether geometrical or arithmetical arguments secured the first proof. From antiquity, only the arithmetical proof for the square has been transmitted. When Plato speaks about the four feet generated by "the diagonal of the diagonal" (*Politicus* 266ab), he certainly alludes to the square. On the other hand, his remark in *Hipp. Maj.* 303b about the sum of two irrational magnitudes being rational may be a hint of the pentagon (Cf. Prop. XIII. 6). Aristotle speaks on several occasions about the "incommensurable diagonal" (as in *Metaphysics* A 2, 983a16), without specifying the kind of diagonal. But when he mentions even and odd numbers in connection with the proof (*Prior Analytics* 41a23 f.), he surely means the square.

The preeminence of the square in our interpretations may simply be a consequence of mathematical instruction and not of historical priority. The square and its diagonal are much more accessible to a nonspecialist audience than the pentagon, and even in mathematical textbooks of today $\sqrt{2}$ stands for the quintessential irrational number.

As proposed by von Fritz [1945], Hippasus the Pythagorean in about 480–450 B.C.E. might have been the discoverer of the phenomenon of incommensurability. Various proofs could have been worked out for both the square and the pentagon.

Observe that at about the same time the "common measure" was popular with architects. All the dimensions of the temple of Zeus in Olympia, completed in 457 B.C.E., are multiples of a unit of 2 feet. Its construction was the most influential architectural undertaking in the first half of the fifth century [Knell 1988, pp. 41–45].

Stage 2

As a replacement of the special arguments in stage 1 the Euclidean algorithm in its geometrical version was introduced as a general method and used by Theodorus (about 440–430?) for the proof of his theorem. The Pythagorean tool of the application of areas, used as demonstrated for $\sqrt{13}$ above, provided incommensurable segments in abundance. This method, however promising, did not solve the general case of $\sqrt{n}$. It got stuck at $\sqrt{19}$ for no apparent reason.

Stage 3

Theaetetus, in the years 400–370, returned to arithmetical methods for $\sqrt{n}$. He proceeded into much more difficult investigations, as are preserved in Euclid's *Elements*, Books X and XIII. Here again special problems with the regular solids may have been the instigators of a general theory. For the regular solids, the diameter of the circumsphere replaced the diagonals of the square and the pentagon. This transition had already been made in XIII.11 for the pentagon. From that time on, the theory of incommensurable magnitudes was accessible only to the most advanced mathematical specialists.

Solid Geometry

···

25.1 The Overall Composition of Book XI

25.2 The Definitions of Book XI

The definitions at the beginning of Book XI are intended to serve for all of the three Books XI–XIII. Accordingly, we find three groups of definitions: Defs. 1–8 determine angles between planes and similar objects. Next it is fixed what is to be understood by similar solid angles (9–11). Pyramids and prisms (12,13) are needed in Book XII, which culminates in the study of cylinders, cones, and spheres.

Def. XI.14.
When, the diameter of a semicircle remaining fixed, the semicircle is carried round and restored again to the same position from which it began to be moved, the figure so comprehended is a sphere.

This sounds more like an artisan's than a mathematician's definition. If a potter were asked to make a spherical vase or a sculptor to carve a solid ball, they would proceed as in this definition. Plato has a similar description of a sphere in his description of the cosmos, but combines it with the definition that a mathematician (of today) would expect:

> And he [the creator] gave to the world the figure which was suitable and also natural … that figure would be suitable which comprehends within itself all other figures. Wherefore he made the world in the form of a globe (sphairoeides), *round as from a lathe, having its extremes in every direction equidistant from the center,* the most perfect and the most like itself of all figures. [*Timaeus* 33a, my italics]

The definitions of cones (18) and cylinders (21) follow the same pattern as Def. 14. Whatever Euclid's motives may have been, these definitions once more convince us that the roots of mathematics lie in the traditions of sculptors, potters, architects, and other artisans.

There is another intrinsic reason for this unusual definition of a sphere. We, being three-dimensional, can look "from above" at a flat circular plate and see its center or reconstruct it as in Prop. III. 1 when it is lost. We would have to be four-dimensional in order to do and see the same things in a solid sphere. On a flat section of the cylindrical part of a column one can recover the center and hence the axis of the column, but the same is impossible for a stone ball.

The remaining, third, group, Defs. 25–28, are about the regular solids and will be quoted in the discussion of Book XIII.

25.3 Foundations of Solid Geometry

There are no foundations of solid geometry in Book XI, at least not in the sense of the axiomatic base of Book I. We may recapitulate from Book I the definition of a plane:

Def. I.7.
A plane surface is a surface which lies evenly with the straight lines on itself.

This is more a description similar to the definitions of points and lines. For solids we have the following analogue:

Def. XI.1.
A solid is that which has length, breadth, and depth.

This, together with Def. XI. 2 about the extremity of a solid, fixes the dimension of solids. The subsequent definitions are technical, like the next one.

Def. XI.3.
A straight line is at right angles to a plane, when it makes right angles with all the straight lines which meet it and are in the plane.

There has been much criticism as to how Euclid tries to derive from these definitions his first propositions, like

Prop. XI.3.
If two planes cut one another, their common section is a straight line.

Modern axiomatic analyses like the one by Hilbert [1899] have shown that there is no alternative to taking this proposition and the preceding one (three noncollinear points are in a unique plane) as axioms for solid geometry. We may conclude that here, as in his arithmetical books, Euclid has preserved a historically earlier stage of mathematical theories. Hence we were right when we said that there is no axiomatic foundation of solid geometry in the proper sense in Book XI. In a broader sense, however, we may speak about foundations because Euclid collects many useful and basic statements about planes, lines, and their angles of inclination in part A of Book XI. As an example take

Prop. XI.14.
Planes to which the same straight line is at right angles will be parallel.

After XI.19 the subject changes to the treatment of solid vertices, or solid angles, as Euclid called them. Propositions 20–23 appear to be a unit, looking like preparatory material for Book XIII in a narrower sense. Neuenschwander [1975, 127] has conjectured that these propositions were stated by Theaetetus as a preparation for his theory of the regular solids, which is preserved in Book XIII.

25.4 The Affinities of Books I and XI

Plato, in his *Republic*, 528 a–d, discusses the state of solid geometry. On the one hand, he praises the "extraordinary attractiveness and charm" (528 d 2) of the subject and observes progress of the investigators in spite of obstacles because of the "inherent charm" of their results. On the other hand, he deplores the state of affairs and speaks of an "absurd neglect" (528 d 9/10) of solid geometry, the reasons for this being (a) the state gives no money for the research in this area and (b) there is no director coordinating the efforts of the investigators (528 b 6–c 4). How would this make sense in the historical situation? We may propose the following solution:

(i) Theaetetus had created the concept of a regular polyhedron and shown the existence of exactly five such polyhedra, certainly a result of "inherent charm" and "extraordinary attractiveness,"

(ii) There did not yet exist a book on the "Elements of solid geometry", so that the results of Theaetetus did not have an elementary foundation consisting of certain simple propositions about the properties of planes etc. comparable to the foundations of plane geometry in section (Aa) of Euclid's Book I.

Under these circumstances, Plato's seemingly contradictory remarks would make perfect sense, and moreover:

(iii) They could be understood as a challenge to write a book on the foundations of solid geometry.

Following the authors quoted below, I assume that such a book was indeed written, that it has, essentially, survived as Euclid's book XI.1–23, 26, and that it was modeled after an existing earlier version of Euclid's Book I. The similarity of Euclid's Books I and XI has been observed by Neuenschwander [1975, p. 118] and Mueller [1981, p. 207], and has been worked out in detail in the paper Artmann [1988]. We direct our attention to some specific details, which show the state of affairs before the introduction of the parallel postulate. The solid geometry in Book XI has no axiomatic foundation like the plane geometry in Book I. This is not what Plato criticizes. On the contrary, it confirms the age of Book XI. In Plato's time mathematics was based on definitions, not on axioms. (See Mueller [1991] for a substantiation.) Parallel straight lines are defined in the very last definition 18 of Book I as lines in the same plane that do not meet. Definition XI. 8 says that parallel planes are those that do not meet. In order to get a better synopsis, we list the propositions of Books I and XI side by side (see Table 25.1).

The most conspicuous analogies between the two books are certainly the respective constructions (11/12) of perpendiculars and (20) the triangular inequality. A closer look at some other propositions is helpful. After 11/12, Book XI goes on to show the uniqueness of the perpendiculars constructed in 11 and 12. This is missing in Book I, but given by Proclus in his commentary on I.17 [Proclus–Morrow p. 244]. Then the two books deviate:

Prop. XI.14:

Planes to which the same straight line is orthogonal will be parallel.

This should be seen in connection with Props. XI. 6 and 8:

Prop. XI.6:

If two straight lines are at right angles to the same plane, the straight lines will be parallel.

TABLE 25.1

Book I		Book XI	
(Aa)		(Aa)	
1–10	foundations	1–5	foundations
		6–10	parallelism and angles for lines and planes
11/12	how to draw lines perpendicular to a given line	11/12	how to draw lines perpendicular to a given plane
		14	a line orthogonal to two planes makes the planes parallel
(Ab)		(Ab)	
13–15	supplementary, vertical angles	15/16,	orthogonality and
		18/19	parallelism for lines and planes
16/17		17	solid analogue to VI.2 (isolated theorem)
(Ac)		(Ac)	
18/19	greater side and angle		
20	triangle inequality for sides of a triangle	20	triangle inequality for angles at a vertex
		21	sum of angles at a vertex
21/22	construction of a triangle from its sides	22/23	construction of a solid vertex from plane angles
23	copying of angles	26	copying of solid angles
24/25			
(Ad)			
26			
(B)			
27/28	equal angles make parallels		
29	parallels make equal angles		
31	construction of parallels		
32	sum of angles in a triangle		

Prop. XI.8:
If two straight lines are parallel, and one of them is at right angles to any plane, the remaining one will also be at right angles to the same plane.

If we, in these three propositions, replace "plane" by "line," then 14 and 6 will coincide and, together with 8, will give us a convenient description of parallels in the (one) plane of Book I. Similarly, the transitivity of parallelism for lines is stated in XI. 9 and I.30.

Proposition XI.15, 16 are specific to solid geometry and cannot have plane analogies. The same is true for XI.18, 19.

Proposition XI.17 is a singularity. It is the solid analogue of VI, 2–10 and is proved by means of VI.2 (proportional segments), which is the basis of all similarity geometry.

With the two triangular inequalities in I.20 and XI.20 we are back to the sequence of Book I. The sum of angles at a solid vertex (XI.21) may correspond to the sum of the angles in a triangle (I.32.) Proposition XI, 22/23 have their exact counterparts in I.21/22. The copying of a plane angle in I.23 corresponds to the copying of a solid angle in XI, 26. At XI.24, another section of Book XI about parallelepipedal solids starts, which has to be compared with section C on parallelograms of Book I. Again the similarities are compelling. In spite of a great number of definitions at the beginnings of both books, neither parallelograms nor parallelepipeds are defined. We leave out a more detailed investigation of the respective propositions because these parts are less important for our purposes. Besides, they seem to be somewhat younger than the preceding material.

Our structural comparison of Books I and XI works in two ways. First, we have seen how Book XI was composed after the pattern of Book I, but with some variations concerning parallels. It is generally assumed that the parallel postulate (5) and the corresponding treatment of parallels in the *Elements*, Props. I. 27–32, were fairly recent at Euclid's time. Hence we may assume that Book XI was modeled after an earlier version of Book I, where parallels were treated in a less sophisticated way. Our second line of investigation will be to recover information from Book XI to find out what that earlier treatment of parallels could have been. Because we have no explicit witnesses, the emerging picture can only be hypothetical.

Nevertheless, it seems to be convincing and may allow us a glimpse at the *Elements* of Leon, which was the textbook of Plato's Academy.

If we look at our schematic representation of the two books (see Table 25.1), we see parallelism in the solid case treated in subsection (Ab), Props. 15–19. The corresponding subsection in Book I has only two remarks about angles and Props. I. 16/17, which we have already recognized in our discussion of Book I as rather late. We conclude that before somebody rearranged the propositions now following I.12, there should have existed a section on parallels and angles and after that a section concerning the triangular inequality. From the plane versions of XI.6, 8, 14 we may even see how parallels were originally constructed. The construction of perpendiculars in Props. XI.11, 12 is the basis for the criterion of parallelism for planes in space, XI.14. In the plane case, Props. I.11, 12 would have served the same purpose. (The more general alternate angles in I.27–29 could be seen as a generalization of right angles.) The pre–Euclidean subsection (Ab) on angles and parallels might then have consisted of the following propositions (we use the numbering of Euclid's *Elements*):

I. 13/14: Angles "on a line" adding up to two right angles.

I. 15: Vertical angles are equal.

I. 27/28/29/31: Alternate angles are equal and make parallels, parallels make equal alternate angles, how to draw parallels.

I. 30: Transitivity of parallelism.

I. 32: The exterior angle of a triangle is equal to the sum of the two interior and opposite angles, and the three interior angles together make two right angles.

At this stage, I.16 would be superfluous, and I.32 could have been used as a basis for a subsection (Ac) on greater relations for lines and angles, consisting of Euclid's Props. I. 18–22 and 24/25. Together with an initial subsection (Aa) consisting of Euclid's Props. I. 1–12, this would have been an attractive (but hypothetical) introduction to plane geometry and fit with the structure of Book XI.

Unfortunately, we have no written evidence of an earlier definition of parallels. There are three key properties of parallel lines (in a fixed plane, as in Book I):

(i) Parallel lines do not meet.

(ii) They make equal (right) angles with (orthogonal) transversals.

(iii) They are equidistant.

All three properties were known by artisans, especially stone masons, from time immemorial. In every wall made from rectangular blocks or bricks one could see parallel lines. Plato acknowledges the strong interrelations of the building trade and geometry in his *Philebus* (56 b–57 d), even if he contrasts practical and scientific geometry in the same place. Maybe even all three properties were stated in the definition. This would be similar to the definition of the diameter of a circle in Def. I. 17, where it is defined as a line through the center that also bisects the circle.

25.5 The Duplication of the Cube

From the rest of Book XI we will pick out just the one theorem Prop. XI. 33, which is closely related to the famous problem of the duplication of the cube.

Prop. XI.33.
Similar parallelepipedal solids are to one another in the triplicate ratio of their corresponding sides.

This proposition is the extension to the solid case of Prop. VI. 20 for similar polygons. In the plane a similarity factor k for segments gives k^2 for areas; now we will have the factor k^3 for the corresponding volumes of similar parallelepipeds.

There is a precisely parallel statement with essentially the same proof in Book VIII about similar solid numbers. We will not discuss similar solid numbers and similar parallelepipedal solids but restrict our attention to cubes in both cases. Except for some very easy modifications, the proofs are the same for the special and the more general situations.

Prop. VIII.19.
Two mean proportional numbers fall between two similar solid numbers; and the solid has to the similar solid the triplicate ratio of that which the corresponding side has to the corresponding side.

Let us first repeat the statements in modern terminology. Let the cubes have sides a and b, so that the respective cube numbers (or volumes) are a^3 and b^3. Then it is said that there are *two* mean proportionals, which is the definition of "in triplicate ratio." This means

$$a^3 : a^2b = a^2b : ab^2 = ab^2 : b^3,$$

the numbers a^2b and ab^2 being the said mean proportionals. In the problem of duplicating a cube, the two numbers $a^3 = 2$ and $b^3 = 1$ are given, and a is sought. Clearly, this amounts to $a = \sqrt[3]{2}$. By transforming the original problem into one about two mean proportionals, it was brought into a version that could be handled by the mainstream methods of Greek mathematics. From Eratosthenes of Cyrene (ca. 250 B.C.E.) we learn that the reduction of the cube duplication to the finding of two mean proportionals was the discovery of Hippocrates of Chios. (For this and many more details of the story see Knorr [1986, esp. p. 23].)

For the proof of Props. XI. 33 and VIII. 19 Euclid uses what one might call a solid gnomon (see Fig. 25.1). We have the two cubes A and B with their respective sides a and b. Let R be the square with side a, and S the rectangle contained by a and b. Then, using Props. VI. 1 and XI. 25/32, we get

$$a : b = R : S = A : C,$$
$$a : b = C : D,$$
$$a : b = D : B.$$

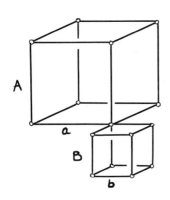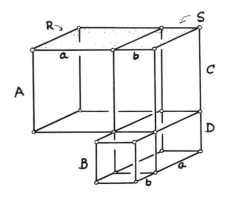

FIGURE 25.1

Hence we have

$$A : C = C : D = D : B,$$

and the two mean proportionals for A and B have been found.

The Greeks could not know that it was hopeless to look for a ruler-and-compass construction of $\sqrt[3]{2}$. Instead of using these "plane" methods, solutions have been found by "solid" methods, that is, spatial curves (Archytas) or conic sections (Menaechmus). For the details see Knorr [1986].

26

CHAPTER

The Origin of Mathematics 13

The Role of Definitions

We have seen how Euclid formulates a definition of the sphere that goes back to the actual process of making one. Plato, in this case more advanced than Euclid, seems to waver between the artisan's and the mathematician's definition when he adds the condition of equidistance from a center. The example shows how a particular mathematical concept changes from a practical and realistic origin toward a more abstract notion that fits better into a developed theory. Usually in the *Elements* we find this mathematically advanced kind of definition. Many of them are explicit descriptions of geometrical objects like "equilateral triangle" or "gnomon," which simply determine certain notions for later use. However, not everything can be defined. Every mathematical theory starts with some undefined concepts. Today these may be in principle no more than "set" and "function" or one of these two, but one cannot start from nothing. In the *Elements* a typical example of this sort is "to measure." We are never told what "measuring" is, in spite of its fundamental importance in Book V for magnitudes and in Book VII for numbers. Somehow one knows what is meant, and it works. We have strong connotations of an everyday procedure that help us to make sense of the concept.

As soon as mathematics develops a little further, another type of definition emerges. When mathematicians discovered that certain segments have no common measure, the notions of commensurability and incommensurability entered the field. This again led to another problem of definition. How should "proportion" be defined, given the new phenomenon? In this way the search for the "right" definition becomes a research problem.

This may sound curious to a philosophically trained reader. After Blaise Pascal (1623–1662) it became firmly established that in mathematics definitions are what are called nominal definitions. The notion "square" is nothing but an abbreviation of the phrase "equilateral and equiangular quadrilateral," and the same is true for any other mathematical concept. In principle one should always be able to replace the shorthand notion by the terms that were used in its definition. In fact, many routine proofs do just this. And more than that: Definitions are never right or wrong, but can only be more or less useful.

What mathematicians mean by a "right" or "good" definition is one that on the one hand fits closely to the more intuitive notion that is under discussion, and on the other hand allows for the fruitful development of the corresponding theory. A paradigmatic example for such a definition is Def.V.5 about the proportionality of magnitudes. A definition of this type has a strong creative component. One cannot know beforehand what will work, and many different approaches may be necessary. Aristotle has a clear understanding of this situation when he speaks about the proof of Prop.VI.1, which is the basis of similarity geometry:

> In mathematics, too, some things would seem to be not easily proved for want of a definition, e.g., that the straight line parallel to the side, which cuts a plane figure [rectangle] divides similarly both the line and the area. But once the definition is stated, the said property is immediately manifest.... [*Topics* 158 b 29–32]

Euclid's definition of similar plane numbers may fall into this same category. A modern example could be the definition of quaternions by Hamilton (1843), who looked for a multiplication in $\mathbb{R}^3$, but found one in $\mathbb{R}^4$ and so created a whole new theory. In this case we are on the borderline of yet another type of definition that mathematicians hold in highest esteem. We might call them creations out of nothing, *creatio ex nihilo*. In the *Elements*, we meet this type in Book XIII about the regular polyhedra. We will see how Theaetetus, knowing three of them from the Pythagoreans, created the concept and classified all examples. This subject is as fascinating today as it was in the time of Plato. Great advances in mathematics are based on definitions of this kind. Newton and Leibniz defined derivatives and integrals; Cantor created infinite cardinal numbers, and Poincaré the fundamental group.

And yet all these definitions are nominal in the strict logical sense. Logical theories are valuable and useful for foundational questions, but in some way they miss what is going on in real-life mathematics. Look, for instance, at theorems that *characterize* a mathematical object. In principle, a characterization should be nothing but the replacement of one definition by another one. Example: A quadrilateral is cyclic if and only if its opposite angles add up to two right angles (cf. Prop. III. 22). Any such characterization will expand our knowledge and reveal, in the important cases, hidden properties and open up new paths of investigations.

Let us add a few words about definitions in the world outside mathematics. Once a mathematical concept is defined, it is and remains fixed in the context of the mathematical theory. For professions other than that of a mathematician's, sometimes exactly the opposite seems to be true. In many cases it is the business of a lawyer or a politician to argue about the meaning of definitions, to expand and twist them until they match one's purposes. The mathematician excludes secondary connotations; the lawyer exploits them. And there is a third person, the poet, who creates them. Writers and poets take care of the living language by creating ever new metaphors and adding fresh connotations to words that have become stereotypes or have worn down over many years of careless usage. The writer Robert Musil (1880–1942), who held a degree in mathematics, says in his great novel "The Man Without Qualities" exactly this: "And so every word demands to be taken literally, or else it would decay to a lie; and yet one cannot take it verbatim, for that would render the world a madhouse" (Musil [1981] p. 749).

27 Euclid Book XII

Volumes by Limits

27.1 The Overall Composition of Book XII

1/2	A: The area of a circle
3–9	B: The volume of a pyramid
10/15	C: Cylinders and cones
16/18	D: Spheres

27.2 The Circle

As a preparation for his main theorem on circles Euclid needs some information on polygons.

Prop. XII.1.

Similar polygons inscribed in circles are to one another as the squares on the diameters.

This is easily reduced to Prop. VI, 20 about the areas of similar polygons by a little geometric manipulation that replaces corresponding sides by the diameters of the circumcircles.

By inscribing successively larger polygons into a circle he approximates the area. This is called the method of exhaustion.

Prop. XII.2.

Circles are to one another as the squares on the diameters.

(In the same way as with polygons in this context he means the area when he says "circle.")

Before sketching Euclid's proof we describe the transition to modern formulas. If two circles have areas A_1, A_2 and diameters d_1, d_2 or (to use the modern convention) radii r_1, r_2, then

$$A_1 : A_2 = d_1^2 : d_2^2 = r_1^2 : r_2^2.$$

Consequently,

$$A_1 : r_1^2 = A_2 : r_2^2,$$

so that the ratio of the area A of a circle to the square r^2 on its radius is constant. This constant is now called π, and we write

$$A : r^2 = \pi, \quad \text{or} \quad A = \pi r^2.$$

The second (and in this context secondary) question is to determine the constant π as precisely as possible. In his *Measurement of a Circle* Archimedes showed that

$$3\frac{10}{71} < \pi < 3\frac{10}{70}.$$

(For details and sources about π the interested reader will find complete information in Berggren et al. [1997].)

The final section of Book XII deals with spheres and establishes

Prop. XII.18.

Spheres are to one another in the triplicate ratio of their respective diameters.

Again we may express this result by saying that if V_1, V_2 are the volumes of two spheres with diameters d_1, d_2 and radii r_1, r_2, then

$$V_1 : V_2 = d_1^3 : d_2^3 = r_1^3 : r_2^3,$$

or, as before, for some constant k,

$$V : r^3 = k.$$

One of the major achievements of Archimedes (*On the Sphere and the Cylinder*, Prop. 34) establishes the following relation between this constant k and π:

$$k = \frac{4}{3}\pi.$$

The Proof of Prop. XII.2

The proof of Prop. XII. 2 is by what has later been called the method of exhaustion. We reproduce Euclid's arguments in modern notation. Let the two circles $\mathcal{K}_1$ and $\mathcal{K}_2$ have areas A_1 and A_2 and diameters d_1 and d_2. There exists a certain area B such that

$$d_1^2 : d_2^2 = A_1 : B,$$

and we have to show that $B = A_2$. Euclid achieves this by excluding $B < A_2$ and $A_2 < B$.

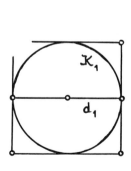
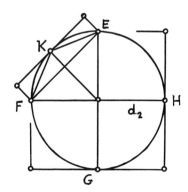

FIGURE 27.1

First, assume $B < A_2$. Inscribe a square $EFGH$ in $\mathcal{K}_2$ and circumscribe another one as indicated in Fig. 27.1. This shows that $\square EFGH > \frac{1}{2}A_2$. In the next step bisect the arcs EF etc. and inscribe an octagon in $\mathcal{K}_2$. Each one of the triangles like $\triangle EFK$ takes away more than half of the area of the remaining segments. At this point Euclid quotes Prop. X. 1 and says, "Thus, by bisecting the remaining circumferences and joining straight lines, and by doing this continually, we shall leave some segments of the circle which will be less than the excess by which the circle $\mathcal{K}_2$ exceeds the area B." Now let $\mathcal{P}_n$ be a polygon such that $B < \mathcal{P}_n$. Inscribe in $\mathcal{K}_1$ a polygon $\mathcal{Q}_n$ similar to $\mathcal{P}_n$. Then, by Prop. XII. 1,

$$\mathcal{Q}_n : \mathcal{P}_n = d_1^2 : d_2^2. \text{ Hence}$$
$$\mathcal{Q}_n : \mathcal{P}_n = A_1 : B.$$

By alternation he derives

$$\mathcal{Q}_n : A_1 = \mathcal{P}_n : B,$$

and this is a contradiction because $\mathcal{Q}_n < A_1$, and $\mathcal{P}_n > B$.

By reducing the case $B > A_2$ to the preceding one the proof is complete.

27.3 The Pyramid

Euclid has defined a pyramid at the beginning of Book XI:

Def. XI.12.

A pyramid is a solid figure, contained by planes, which is constructed from one plane to one point.

Clearly, the Greeks knew the great pyramids of Egypt. Just imagine that for them these pyramids were as old as for us a building constructed by the Roman emperor Augustus some 2000 years ago. One might think of an Egyptian origin of the word "pyramid" itself, but the Greek dictionary gives a more plausible explanation by the way of *pyra* = funeral pile (associated to *pyr* = five) with the extended meaning of a holy burying place. This is exactly what the Egyptian pyramids were.

From elementary geometry we know the formula for the volume P of a pyramid with base b and altitude h to be $P = \frac{1}{3}bh$. (See, for instance, the high-school text by Jacobs [1987], who states it without proof.) Euclid proves this result and expresses it in his way without a formula. We will sketch his theory and see how he realizes various aspects of $P = \frac{1}{3}bh$. (For the proofs see Heaths's text.) His general procedure is, even if more complicated, similar to his treatment of the circle. The first step is to find a solid replacement for the polygons inscribed in a circle (see Fig. 27.2).

Prop. XII.3.

Any pyramid which has a triangular base is divided into two pyramids equal and similar to one another, similar to the whole and having triangular bases, and into two equal prisms; and the two prisms are greater than the half of the whole pyramid.

The second step provides us with an analogy to Prop. XII. 1 by looking at the proportions of the approximating solids. Let P_1, P_2 be two triangular pyramids with bases b_1, b_2 and equal altitudes, and let W_i be the combined volumes of the two prisms in P_i.

Prop. XII.4.
[abbreviated] $W_1 : W_2 = b_1 : b_2$.

Step 3 is the proof by exhaustion of the main result.

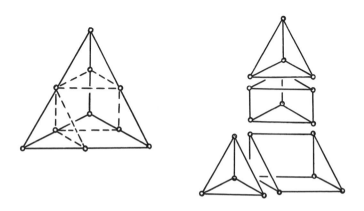

FIGURE 27.2

Prop. XII.5.

Pyramids which are of the same height and have triangular bases are to one another as the bases.

The proof of XII.5 follows the pattern of XII.2. From XII.3 we have the subdivision of a pyramid P_0 into two prisms (more than its half) and two smaller pyramids P_1, P_2. Subdivide these as before and iterate, and so on.

What follows is a technical extension of the preceding result for pyramids with triangular bases.

Prop. XII.6.

Pyramids which are of the same height and have polygonal bases are to one another as the bases.

Let us see how close we are to the formula. For pyramids with equal heights we know that

$$P_1 : P_2 = b_1 : b_2.$$

If we ignore dimensions (something that Euclid could not do) and regard P_i, b_i as real numbers, we could alternate to

$$P_1 : b_1 = P_2 : b_2 = k, \text{ a constant.}$$

Even if Euclid could not calculate as we did, he determines k in his next step. (Fig. 27.3).

Prop. XII.7.

Any prism which has a triangular base is divided into three pyramids equal to one another which have triangular bases.

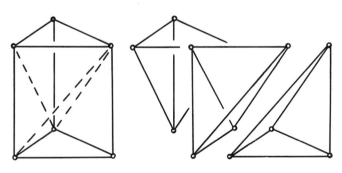

FIGURE 27.3

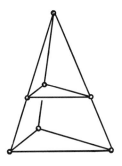

FIGURE 27.4

This gives us $k = \frac{1}{3}$. We still do not yet know how the volume changes with different heights. If one cuts a pyramid by a plane parallel to the base, the problem is that the size of the base changes together with the size of the height. (Fig. 27.4) In his next proposition, Euclid solves this problem in a roundabout way. He extends Prop. XI. 33 about similar parallelepipeds to pyramids.

Prop. XII.8.
Similar pyramids which have triangular bases are in the triplicate ratio of their corresponding sides.

This together with the last proposition solves the problem of the volume of pyramids completely.

Prop. XII.9.
Let P_1, P_2 be equal pyramids with triangular bases b_1, b_2 and altitudes h_1, h_2. Then

$$P_1 = P_2 \quad \Leftrightarrow \quad b_1 : b_2 = h_2 : h_1.$$

As with triangles, Euclid never measures volumes. He thinks geometrically and does not measure volumes in our sense, that is, he does not assign a number to a solid as its volume. On the other hand, one could, given the above information and starting from a unit cube, derive the formula $P = \frac{1}{3}bh$.

27.4   Cylinders, Cones, and Spheres

Euclid's treatment of cylinders, cones, and spheres follows the pattern established for the circle and the pyramid. The geometry gets more complicated, but the method of exhaustion is basically the same. Only in Props. XII, 13/14 does a new element enter the proofs. In Prop. XII. 14 he treats cylinders cut by planes parallel to the base (that is, of different heights) and employs the definition Def. V. 5 of proportionality in a way similar to Prop. VI. 1. Thus he gets

Prop. XII.14.
Cones and cylinders on equal bases are to one another as their heights.

One wonders why he does not state a similar result for prisms and pyramids. At this point one could speculate about the contributions of various pre-Euclidean authors to the text, but we will leave this to the specialists.

28

The Origin of Mathematics 14

The Taming of the Infinite

....

Using finite means for mastering the infinite is a hallmark of mathematics. Here again Euclid did ground–breaking work. We will list a series of his examples, all of which were discussed in greater detail at their respective places.

First example: The theory of parallels

Parallels are defined as nonintersecting lines. In principle, then, we have to go an infinite way in order to find out whether two straight lines in a plane meet. Euclid proposes a finite solution to the problem: Just check alternating angles! (In fact, the parallel postulate is the first decisive step toward this solution.) Observe that the definition of parallels as equidistant lines would also involve the infinite.

Second example: Tangents to a circle

Nowadays, tangents to curves are defined using limits. For the special case of a circle the problem is easier. In Book III, Def. 2 Euclid defines the tangent to a circle as a line touching the circle in just one point. More than a hundred years before Euclid, the philosopher Protagoras of Abdera (ca. 480–400) had opposed the very notion of a tangent. He claimed that it was impossible for a line to meet a circle in exactly one point. Probably he thought of something like an infinitely small segment being common to line and circle. Whatever the precise belief of Protagoras's had been, Euclid transformed the more "intuitive" definition into the following statement: The line g touches the circle with center C in the point P if and only if the line CP is orthogonal to g. There is no need to determine a limit in order to find the tangent to a circle (Props. 16 – 19 of Book III).

Third example: The proportionality of line segments

For the proportionality of four magnitudes a, b, c, d we have the sophisticated Def. V. 5 of Book V: For all numbers n, m, if $na > mb$, then $nc > md$, and so on. If we don't find anything better, we have to check an infinity of cases. Again Euclid presents a "finite" criterion, if only for the proportionality of lines (segments). Definition V, 5 is used in Prop. VI. 1 in order to establish a proportionality between two different domains of magnitudes, lines, and [areas of] rectangles (or triangles). The immediate consequence of Prop. VI. 1 is the theorem on proportional segments, Prop. VI. 2 with its simple condition for proportionality of a parallel cutting the side of a triangle.

    A second geometric criterion, as useful as the first, is established in Props. VI. 14/16 for the proportionality of the lines a, b, c, d in terms of the rectangles $\square$ (a, d) and $\square$ (b, c) contained by the respective lines.

$$a : b = c : d \iff \square\, (a, d) = \square\, (b, c),$$

the latter equality meaning "of equal content." The equality of areas, which can be handled effectively by the methods of Books I and II, replaces the more involved Def. V. 5 which in modern terms translates into taking limits. This second criterion is again based directly on Prop. VI, 1. (It is once more expanded in Props. VI. 24/26.) Euclid is a master in handling proportions for lines, and he uses VI. 1 whenever possible to reduce proportions for magnitudes of higher dimensions (e.g., areas) to the case of lines. A typical example for this procedure is the proof of Prop. VI. 19.

Fourth example: Prime numbers

Whenever we have a finite number of primes, these will not be all of them. In one simple step, executed on a finite number, Euclid catches the infinite.

Fifth example: Incommensurability

This is more delicate. Commensurability is defined in Def. X. 1 by the existence of a common measure. This is the finite side of the problem. Proposition X. 2 gives a criterion for commensurability in terms of the Euclidean algorithm (anthyphairesis) for magnitudes, and Props. X. 5/6 another one in terms of (natural) numbers.

The only way to find out that the anthyphairesis algorithm does not terminate seems to be to establish periodicity. Periodicity is a finite criterion for infinite repetition and hence incommensurability. There is no proof of this type in the *Elements*. The second way to tackle the problem is to use the criterion "as number to number" and to produce a contradiction, as done in X. 117. In principle, this method proceeds thus: Assume $a : b = k : n$ and let $k : n$ be in lowest terms (minimi). Use some geometric argument in order to derive numbers r, s smaller than k, n with $a : b = r : s$, a contradiction to "in lowest terms." One step is enough in order to solve an infinite problem.

This method was made popular by Fermat (1601–1655) and has historically become known under the name "infinite descent." Its base is the well-ordering of the natural numbers: *Every nonempty subset of the set of natural numbers* $\mathbb{N}$ *has a smallest element*. This is equivalent to the method of mathematical induction, which covers infinitely many cases with a finite amount of information compressed into one step.

Sixth example: Exhaustion

Here we meet limits in an explicit way. The sequence of (areas of) polygons approaching the circle is used to prove a theorem about circles. The existence of the limit, the area of the circle, is no problem for Euclid. The central fact is that he can make the difference between the circle and the approaching polygons as small as wanted, and this is nothing else than the modern definition of convergence. By going out far enough in the finite realm one can prove statements about the infinite; that is the message of this example.

29

Euclid Book XIII

Regular Polyhedra

....

29.1 The Overall Composition of Book XIII

1–6	A: Division in extreme and mean ratio
7–12	B: About the pentagon, and lemmas for part C
13–18	C: Regular polyhedra

Parts A and B are mainly preparatory for part C, but some of the theorems in part B are of independent interest. We will discuss the latter ones in this section. The history of the regular polyhedra from Hippasus to finite simple groups will be the subject of the next section, "Symmetry through the ages."

29.2 Division in Extreme and Mean Ratio and the Pentagon

In Def. VI.3 a segment PQ is said to have been cut in extreme and mean ratio by a point R (with greater segment PR) if

$$PQ : PR = PR : RQ.$$

(In the first half of the nineteenth century this was given the name "the golden section.") We will abbreviate it by EM ratio. Given the segment PQ, the point R has already been found twice: in disguise in Prop. II. 11 and explicitly in Prop. VI. 30. Here we are confronted with a third construction, again without motivation as in Prop. II. 11. We abbreviate Euclid's statements of Props. XIII, $1/2$. Let the points S, P, Q, R lie on a line as shown in Fig. 29.1 and let $PQ = 2PS$. Then we have (XIII. $1/2$) that

$$PRQ \quad \text{are in} \quad \text{EM ratio} \quad \Leftrightarrow \quad \square SR = 5 \,\square\, PS.$$

Since we can easily construct a square equal to $5 \,\square PS$, we have another construction of the EM ratio. Props. XIII. $3/4$ are variations of $1/2$. All of the propositions have proofs that resemble the methods of Book II.

The next proposition is about the iteration of the EM ratio as we have met it in our speculations of the Fibonacci sequence in Chapter 24, see Fig. 29.1(b). It could be read off geometrically from an iterated pentagram, but here it is proved directly from the definition by

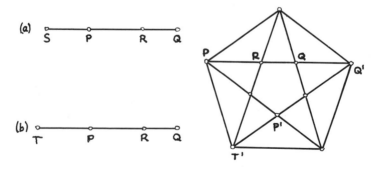

FIGURE 29.1

manipulating rectangles in a way that looks more like algebra than geometry.

Prop. XIII.5

(See Fig. 29.1 (b)): *If PRQ are in EM ratio and TP = PR, then QPT are in EM ratio.*

It should be stressed once more that the iteration here is going "outward" to bigger segments as in the Fibonacci sequence. We have seen how the iteration "inward" in the pentagram or for the EM ratio will immediately lead to the respective incommensurability statements, either in the arithmetical or in the geometrical version. (Euclid reveals the motivation behind the pentagon/pentagram only in Prop. XIII. 8, as we have discussed in our speculations about the pentagon in Book IV.) Thus it is quite natural to find a theorem about the incommensurability of the diagonal and side of the regular pentagon at exactly this place, even if it is stated just for the EM ratio. However, what we find is not the simple incommensurability statement, but a refined "now I know better" version employing the vocabulary of Book X.

Prop. XIII.6.

If a (rational) straight line is cut in extreme and mean ratio, each of the segments is an irrational straight line, the one called apotome.

The proof uses the fact that by Prop. XIII. 1 the respective segments can be expressed as rational combinations of $\sqrt{5}$, and the definition of an "apotome" in Prop. X. 73, which expresses just this in slightly more general terms. There has been a debate among philologists as to whether this proposition was in fact contained in Euclid's original *Elements* or whether it is a later interpolation. (See Heath on XIII. 6.) In agreement with Eva Sachs [1917], pp. 112–116, I am convinced that it is genuine. In fact, Theaetetus, who is generally considered to be the author of the original versions of Books X and XIII, may have replaced an older statement by a more modern one as indicated above and thus have created the philological problems. From Prop. XIII. 6 we will pass directly to XIII. 11, because there is a considerable and significant shift of interest between these related propositions, which again may show the handwriting of Theaetetus. Whereas XIII. 6 is about the diagonal and the side of the regular pentagon, XIII. 11 is about the diameter of the circumcircle and the side of the pentagon. In (regular) polygons with more than 5 sides, there are (many) different diagonals, and in triangles there are none at all. On the other hand, the diameter (or radius) of the circumcircle is a standard magnitude for all of them, which can be compared with the side. In fact, exactly this is done for the equilateral triangle in Prop. XIII. 12. Thus Props. XIII. 11/12 are decidedly more abstract and general in their setting than Prop. XIII. 6, which is confined to the specific situation of the pentagon. Moreover, Props. XIII. 11/12 lead directly to the subsequent study of the interrelations between the diameters of the circumspheres and the edges of the regular polyhedra.

Prop. XIII.11.

If an equilateral pentagon is inscribed in a circle having a rational diameter, the side of the pentagon is an irrational straight line, the one called minor.

In part 5 of Chapter 13, "Polygons after Euclid," we calculated the side and diagonal of a regular pentagon inscribed in the unit circle. Adjusting the formula for a circle with diameter d, we get for the side

$$f = d\sqrt{10 - 2\sqrt{5}}.$$

Euclid finds and expresses this in his geometrical terms. The interested reader will find detailed analyses of the proof either in Heath's comments or in Taisbak [1982], 9–15. Taisbak takes this proof as his starting point for the investigation of Book X. Together with Mueller [1981], 260–263, Taisbak proposes to see this and the calculation of the edge of the icosahedron in Prop. XIII. 16 as the main motivations for the attempted classification of irrational segments in Book X.

29.3   About the Decagon

The two Props. XIII. 9/10 are about the decagon and its relations to the pentagon and the hexagon. (All polygons are understood to

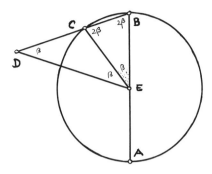

FIGURE 29.2

be regular.) After studying the two propositions we will find very similar ideas expressed in a famous Greek temple, whose architect lived at about the same time as Theaetetus.

Prop. XIII.9.
If the side of the hexagon and that of the decagon inscribed in the same circle are added together, the whole straight line has been cut in extreme and mean ratio, and its greater segment is the side of the hexagon.

For his proof, Euclid essentially uses the triangle of Prop. IV. 10 (Fig. 29.2). In IV. 10 he avoided proportions, but in effect started from a line in EM ratio and had to find the relevant angles. This time he knows the central angle of the decagon ($\beta = 36°$) and the sizes of the respective segments and has to show that they are in EM ratio. Hence Prop. XIII 9 is a sort of converse to Prop. IV 10. The two propositions combined are something like a special case of the most important Props. VI 4/5 about proportional segments and angles. As in Prop. IV 10, for the proof Euclid uses various isosceles triangles. Because he now may use similarity arguments, he gets the desired result from $\triangle EBC \approx \triangle DEB$.

Prop. XIII.10.
If an equilateral pentagon is inscribed in a circle, the side of the pentagon is equal in square to that of the hexagon and that of the decagon inscribed in the same circle.

In other words, the side of the pentagon is the hypotenuse of a right triangle that has the sides of the hexagon and of the decagon as its legs—a really unexpected and beautiful insight! Following Sachs

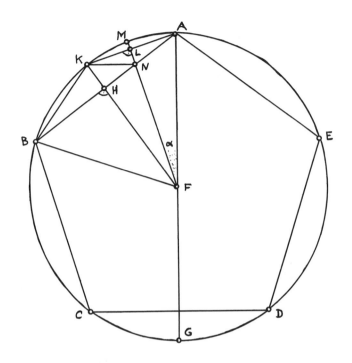

FIGURE 29.3

[1917] and Neuenschwander [1975], Mueller [1981], pp. 258/59 shows how XIII. 10 could have been discovered in the course of the investigation of the icosahedron, a very plausible suggestion. In fact, XIII. 10 is used during the construction of the icosahedron in XIII. 16. (It is common in mathematics that certain parts of complicated proofs become of independent interest, and afterwards one wonders how anybody could have thought of them.)

On the other hand, Taisbak [1982] presents a relatively simple derivation of Prop. XIII. 10 within the confines of plane geometry. Later on, Prop. XIII. 10 has become important in applied mathematics, too. Ptolemy [1974, *Almagest* I, 10] uses it as a valuable tool in his calculations of the tables of chords (that is, the trigonometric functions).

The Proof of Prop. XIII.10

As usual, we emphasize the main steps of the proof, abbreviate a little, and use modern notation; in particular, numerical values are used for the sizes of the angles.

Step 1.

(See Figs. 29.3 and 29.4) The regular pentagon $ABCDE$ is inscribed in the circle with center F. The arc AB is bisected at K and the arc AK again at M. Thus AK is the arc of the decagon and AM the arc of the icosagon (20–gon). The points are labeled as in Fig. 29.3, and the respective segments are drawn. Note that the angle $\alpha = \angle AFM$, of which all other relevant angles will turn out to be multiples, has measure $18°$.

Step 2.

To show that the triangles $\triangle ABF$ and $\triangle BFN$ are similar. The isosceles triangle $\triangle ABF$ has $\angle AFB = 72°$, and the two other angles measure $54°$.

In $\triangle BFN$ we have $\angle BFN = \angle BFM = 72° - 18° = 54°$. Angle $\angle NBF$ equals $54°$ as in the first triangle; hence $\angle BNF$ must equal $72°$, and the two triangles are equiangular. This implies $AB : BF = BF : BN$ by Prop. VI. 4, and hence $\square(AB, BN) = \square BF$ by Prop. VI. 17.

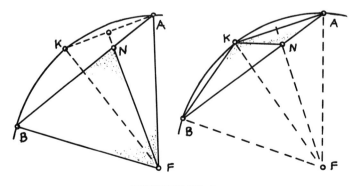

FIGURE 29.4

Step 3.

To show that the triangles $\triangle KBA$ and $\triangle NKA$ are similar. Because FM is the perpendicular bisector of AK, we have $AN = KN$ and hence $\angle AKN = \angle KAN$. But the latter angle is common to the two isosceles triangles. Thus $\triangle KBA$ and $\triangle NKA$ are equiangular, and as above we have $BA : AK = AK : AN$; hence $\square(AB, AN) = \square AK$.

Step 4.

Now observe that for the side AB of the pentagon, $\square AB = \square(AB, AN) + \square(AB, NB)$. Inserting what we know from steps 2 and 3 for the two rectangles, we get the assertion

$$\square AB = \square BF + \square AK,$$

where BF is the side of the hexagon and AK the side of the decagon.

The Icosagon at Delphi

The icosagon, or regular 20-gon, is a combination of two of the most prominent figures in plane geometry, the square and the regular pentagon (see Fig. 29.5). Euclid does not mention the icosagon, but we have seen that its central angle $\alpha = 18°$ played a hidden, but essential, role in the proof of Prop. XIII. 10. The icosagon is the

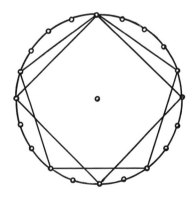

FIGURE 29.5

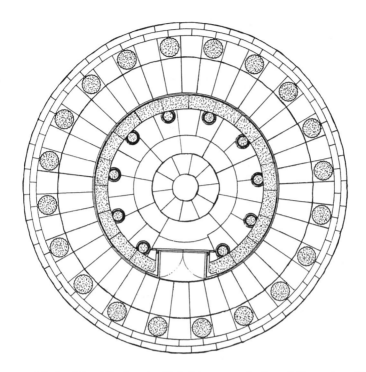

FIGURE 29.6 The Tholos in Delphi, ground plan (Diameter 50 feet, that is, about 14.75m) After Bousquet [1993], p. 294 and Charbonneaux [1925].

mathematical foundation of one of the most famous Greek temples, the Tholos (round temple) in the sanctuary of Athena at Delphi.

The 20 outer columns of the Tholos form a perfect regular icosagon in a circle (Figs. 29.6 and 29.7). We know that the Tholos was built in about 380–370 B.C.E. by the architect Theodorus of Phokaia, who even wrote a book about its construction. Unfortunately, this book has been lost, but we can easily see the architect's fascination with geometry from what remains of the Tholos. Each one of the 20 columns has 20 grooves, so that its cross section is again a regular icosagon, and all the columns are 20 feet high. The execution of the masonry is as perfect as the geometry.

At the same time, about 380 B.C.E., Theodorus of Phokaia realized in the architecture of the Tholos what Theaetetus of Athens investigated in his geometrical theories. In their respective fields they

FIGURE 29.7   The Tholos in Delphi

each created what we justly may call some of the finest pieces of the cultural heritage of the Greeks. Did Theodorus of Phokaia and Theaetetus of Athens know each other? It seems likely, but we will never know for sure. We see their works, and they are wonderful enough.

29.4   The Regular Solids

The five regular solids—the tetrahedron, the octahedron, the cube, the icosahedron, and the dodecahedron—are the main subject of Book XIII (see Fig. 29.8). Each of the solids is constructed and inscribed on a given sphere. At the end Euclid shows that there cannot be more regular solids than the constructed ones. In addition to this he investigates the relations between the length of an edge of the solid and that of the diameter of the circumscribing sphere, using some of the notations of Book X.

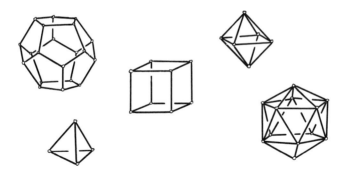

FIGURE 29.8   The five regular solids

We will restrict our attention to the relevant definitions from Book XI, the construction of the dodecahedron and the final remark that no other regular solid can be constructed. A detailed study of the history of the regular solids will follow in the next section, "Symmetry Through the Ages."

The definitions from Book XI

Def. XI.12.
A pyramid is a solid figure, contained by planes, which is constructed from one plane to one point.

Pyramids in general were treated in Book XII. In Book XIII, a "pyramid" is always meant to be a regular tetrahedron, that is, a pyramid that has four equal equilateral triangles as its base and the three other faces.

Def. XI.25.
A cube is a solid figure contained by six equal squares.

Def. XI.26.
An octahedron is a solid figure contained by eight equal and equilateral triangles.

Def. XI.27.
An icosahedron is a solid figure contained by twenty equal and equilateral triangles.

Def. XI.28.
A dodecahedron is a solid figure contained by twelve equal, equilateral and equiangular pentagons.

The Construction of the Dodecahedron

For the construction of the dodecahedron, Euclid starts with a cube and constructs what can be called a "roof of a house" over each of its faces in such a way that twelve regular pentagons appear. This construction suggests itself if one knows certain crystals of the mineral pyrite (see Fig. 30.1). Because the edges of the cube reappear as the diagonals of the pentagons, the essential point for the construction is to know the ratio of the diagonal d and the side f of a regular pentagon; or in other words, the division in extreme and mean ratio.

Step 1. The construction

For the labeling of the various points see Fig. 29.9. We start from a cube *ABCD* etc., bisect the edges *EB* in *N* and *FD* in *O* and join *N* and *O*. Let *P* be the center of *NO* and the segments *NRP* and *OSP* in EM ratio, with greater segments *PR* and *PS*. Proceed similarly on the other faces of the cube as indicated in Fig. 29.9.

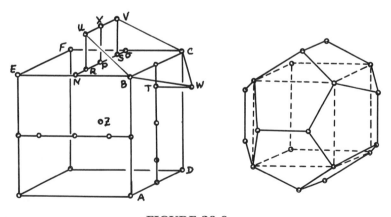

FIGURE 29.9

Let the segments RU, SV, and TW be orthogonal to the respective faces of the cube and each of them equal to RP. Joint the five points $BWCVU$ to obtain a (possibly nonplanar) pentagon. Proceed similarly at the other faces of the cube.

Step 2. The pentagon obtained is equilateral

First observe that 2 PR and 2 RN are in EM ratio (equal to $PR : RN$). Hence the segment $UV = RP$ is in EM ratio to the edge $BC = NO$ of the cube.

Following Euclid we will show $UV = UB$. Applying Pythagoras's theorem twice, we get (using the conventional BU^2)

$$BU^2 = RU^2 + BR^2 = RU^2 + BN^2 + NR^2.$$

Now, BN equals PN and RU equals PR and PRN are in EM ratio with greater segment PR. By Prop. XIII. 4 this implies

$$BN^2 + NR^2 = PN^2 + NR^2 = 3PR^2;$$

hence

$$BU^2 = RU^2 + 3PR^2 = 4PR^2,$$

and $BU = 2PR = UV$ follows. Obviously, the other edges of the pentagon are of the same length.

Step 3. The pentagon is planar

In order to show that the points $BWCVU$ are in one plane, it suffices to show that the line joining the midpoint X of UV and the point W passes through the midpoint of the line BC. This is done by Euclid using Prop. VI. 32 as his principal tool.

Step 4. The pentagon is equiangular

Euclid first shows the equality of three angles by using congruent triangles, and then the assertion is established by Prop. XIII. 7.

Step 5. To inscribe the dodecahedron in a sphere

For each of the points of the original cube, which are also points of the dodecahedron, the distance of the point from the center Z of the cube is determined by, e.g.,

$$ZB^2 = 3BN^2.$$

It has to be shown, e.g., that

$$ZU^2 = 3BN^2.$$

Again Euclid employs the theorem of Pythagoras and Prop. XIII. 4 in order to prove this relation.

For the completion of his study of the dodecahedron Euclid classifies the edge of the dodecahedron as an apotome in the terminology of Book X.

29.5 The Classification of the Regular Solids

Euclid's treatment of the regular polyhedra is especially important for the history of mathematics because it contains the first example of a major classification theorem. Such theorems start with a definition (or axiomatic description) and end with a list of objects satisfying the description. There is no official definition of a regular solid in the list of definitions at the beginning of Book XI. However, we find one in the very last theorem of the *Elements*, which asserts the completeness of the preceding list of polyhedra.

Prop. XIII.18a.
No other figure, besides the said five figures, can be constructed which is contained by equilateral and equiangular figures equal to each other.

Euclid apparently regards a solid as regular if its faces are congruent regular polygons. This is essentially the modern definition except for two omissions: the requirement that the polyhedron be convex, and the specification that the solid angles (vertex figures)

FIGURE 29.10

of the polyhedron be congruent. Euclid probably took the first of these conditions for granted, but overlooked the existence of the kinds of convex polyhedra illustrated in Fig. 29.10, as as been pointed out by Freudenthal and van der Waerden [1946]. Instead of requiring the equality of the vertex figures one might add the condition that the solid have a circumsphere, which is implicit in Euclid's Props. XIII. 13–17. Plato seems to have such a description of a regular solid in mind when he speaks of a solid "that gives rise to a subdivision of the (circum-)sphere into equal and similar parts" (*Timaeus* 55a). Similarly, Proclus speaks of the problem of inscribing in a sphere "polyhedra with equal sides and angles and composed of similar faces" (Proclus–Morrow p. 158), which again makes the existence of a circumsphere part of the definition. Adding this condition to Prop. XIII. 18a would make the theorem true but complicate its proof. If we concede Euclid the equality of the vertex figures, then his proof is valid and easy. Looking at the angles of regular polygons and remembering Prop. XI. 21, which says that any solid angle is contained by plane angles less than 4 right angles, the proof can in effect be read off from Fig. 29.11.

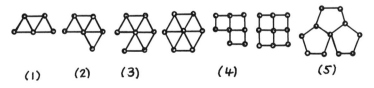

FIGURE 29.11   (1) tetrahedron, (2) octahedron, (3) icosahedron, (4) cube, (5) dodecahedron

Proclus states that the aim of the *Elements* is "both to furnish the learner with an introduction to the science as a whole and to present the construction of the several cosmic figures [the regular solids]" (Proclus–Morrow p. 59).

This characterization is too humble. The *Elements* contains much more than a first introduction to mathematics, and, as we have seen, the study of the regular solids is only one of several highlights.

30

CHAPTER

The Origin of Mathematics 15

Symmetry Through the Ages

···

From the time of the Greeks, people have been fascinated by the regular polyhedra. They provide us with one of the first complete mathematical theories: a general definition together with a complete classification of all the objects satisfying the definition. In this section the most important steps are presented in the development of a subject that goes back to the very beginnings of mathematics and is still alive today. Hippasus provided the first significant example, and Theaetetus created the mathematical theory. Pacioli revived the subject after it had lain dormant for about a thousand years. Felix Klein replaced the polyhedra by their symmetry groups and opened vast new areas of research. One path leads into function theory and algebraic geometry, while the other starts with the group of rotations of the dodecahedron and goes on to simple groups.

Outside of mathematics there flows another stream of ideas related to the regular polyhedra. Plato associated them with the elements. From Roman times we have numerous dodecahedra whose purpose remains a mystery. Pacioli, while keeping the contents of his book strictly mathematical, is so enthusiastic about the polyhedra, and especially the golden section, that he may well have inspired the German art historian Zeising in the middle of the last century. Zeising claimed to have found the key to all secrets of beauty in the proportion of the golden section (or EM ratio).

The beauty of the regular solids does not reside in their physical appearance; it lies hidden in the realm of mathematical thought. The interplay of the general concept of regularity and its realization in exactly five solids can be grasped only by mathematics. Plato was the first to understand this. The participation of special objects in one general idea lies at the center of his philosophy.

30.1 Nature

Dodecahedra appear naturally. The mineral pyrite (chemical formula FeS_2) can crystallize in any of three shapes: as cubes, as octahedra, and as almost regular dodecahedra (Fig. 30.1). Pyrite was well known to prehistoric man, first as a common fire starter, and later on as an important iron ore. When hit with a stone, pyrite gives relatively long-lasting sparks, which could be caught by tinder. This is well known to archeologists; see, for instance, [Ebert]. The association with fire is preserved in the Greek word *pyr* = fire. We will see below that the Pythagorean Hippasus of Metapontum (about 470–450 B.C.E.) was most likely the first mathematician who studied the dodecahedron. Aristotle mentions in his *Metaphysics* (A3, 984a7) that Hippasus regarded fire as the first principle. This may be another reference, however vague, to the connection between fire and the dodecahedron.

30.2 Art

About 390 neolithic carved stone balls of fist size dating from before or about 2000 B.C.E. have been found in Scotland. All of the

FIGURE 30.1 Crystals of pyrite (diameter about 1 cm)

five regular solids appear in these decorations, the dodecahedron
on one specimen in the Museum of Edinburgh. A scientific description in Ritchie and Ritchie [1981] calls them a "... relatively short
lived and peculiar Scottish phenomenon." Very good photographs
are to be found in Critchlow [1979], who is, however, otherwise very
speculative. A picture with five "regular" stone balls is reproduced
in *Mathematics Teaching*, Vol. 110 (March, 1985), p. 56.

Bronze dodecahedra were popular in Roman Imperial times.
More than 70 of them, mostly from the third or fourth century C.E.,
can be seen in various museums in Western Europe. Most of them
come from the northeastern part of France, from Switzerland, or
from the Roman parts of Germany along the Rhine river. A detailed
account from the archeological point of view and a complete list of
the pieces in various museums have been given by Nouwen [1993]
(in Dutch).

These dodecahedra are generally of about fist size or a little
smaller, and in the vast majority of cases look similar to the one
in Fig. 30.2, which is displayed in the Landes–Museum in Mainz
(Germany).

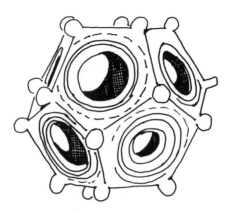

FIGURE 30.2

They are hollow with circular holes of differing sizes in the faces and with knobs at their vertices. Many of them have impressed rings around the holes. There are a few exceptions to the rule. The first one was described by Lindemann (of π) [1896]. He reports on a do-decahedron cut from soapstone with unintelligible markings on the faces, which was found in northern Italy. It is supposedly of Etruscan origin, dating from about 500 B.C.E.. The precise circumstances of its excavation have been lost, and we have no way of being certain of Lindemann's claims.

There are, however, two Etruscan bronze dodecahedra in the Museum of Antiquities in Perugia (Italy). They are mounted on bronze sticks and are without any further decorations of mathematical nature.

Another singular example was excavated quite recently in Geneva (Switzerland) (Cervi–Brunier [1985]). Its edge length is about 1.5cm. Its faces are made of silver and are inscribed with the names of the signs of the zodiac. The core is solid lead. For a picture see Artmann [1993].

The meaning of these dodecahedra is an open problem in archeology. Various hypotheses have been proposed, some of them quite fanciful, like a "surveying instrument," but nothing is known for certain. I think that two conjectures look promising: the association of the dodecahedron with fire as mentioned above, and some connections to the zodiac. Deonna [1954] quotes Plutarch as saying that the

FIGURE 30.3 Roman icosahedron (Rheinisches Landesmuseum Bonn Inv. 53.356.) Reproduction with kind permission of the museum.

dodecahedron is a sort of image of the zodiac or the year because all three have twelve parts.

Apart from the dodecahedra from western Europe, we have a few small icosahedra from Hellenistic or Roman Egypt that have the numbers 1–20 inscribed on their faces and were apparently used as dice.

Together with 77 dodecahedra, Nouwen [1993] shows one singular icosahedron. Like the dodecahedra, it is hollow and made from bronze. It weighs 465g and its overall diameter is about 8cm (Fig. 30.3).

The icosahedron was excavated in 1953 in the village of Arloff (some 30 km SW of Bonn, Germany). At that time, it was erroneously classified as a dodecahedron and put into storage in the Rheinisches Landesmuseum in Bonn. There it sat in the basement until recently, when Dr. Ursula Heimberg, on the staff of the museum, had another look and discovered that it was not at all a dodecahedron, but an icosahedron. As Nouwen says, this icosahedron is even more mysterious than the dodecahedra. Nobody knows what its use or purpose was.

Dodecahedra may have been inspired by crystals and may have a meaning outside of mathematics in connection with the zodiac.

Icosahedra seem to be different. They were discovered "inside" mathematics and have specific significance only as members of the class of regular polyhedra. It is hard to imagine anything other than a mathematical origin of the icosahedral form of the object from Arloff. That is, at least a little knowledge of Euclid's *Elements* must have diffused to the northwestern provincial regions of the Roman Empire. Mathematical knowledge must have been much more widespread during imperial Roman times than has been supposed until now.

After a long interlude, the subject of the regular solids was taken up again in Italy. Because mathematical instruction from late medieval times onward was based mainly on Euclid (as it had been in antiquity), knowledge of the five regular solids was widespread in about 1400 C.E..

The famous Renaissance painter Piero della Francesca (ca1420–1492), wrote a treatise about "the five solids." His student Luca Pacioli (1445–1515) translated, edited, and enlarged this book and published it in 1509 under the title *Divina Proportione*, claiming that this was "... a work essential to all open spirits curious for knowledge."[1] The pictures in this book were provided by his friend Leonardo da Vinci. Another author from the Renaissance, Wenzel Jamnitzer [1568/1973], drew more than fifty fine pictures of the regular solids (some together with their circumspheres) in his *Perspectiva Corporum Regularium*.

Pacioli published some more mathematical books "... from compassion with the ignorant." All of them are rather elementary. But the point is that he and the other mathematical authors of his generation revived Greek mathematics and made it accessible to their students, the famous Italian Renaissance algebraists. The mathematics in the *Divina Proportione* consists essentially of excerpts from Euclid. Pacioli stresses the mathematical importance of the golden section (divina proportione, in his words, or EM ratio, in Greek words), but strictly confined to mathematics. Here are some of his chapter headings.

---

[1] I wish to thank Peter Hilton for the translation of this and all other quotations from Pacioli.

XXII. "Of its thirteenth and most important elaboration. How without the knowledge of this proportion the construction of the regular pentagon is impossible. How Euclid in his proofs applies only that which precedes and not that which follows."

XXIV. "How the stated elaborations contribute to the classification of all regular bodies and those dependent on them. Why these five bodies are called regular."

XXV. "That it is impossible for there to be more than five regular bodies and why. That it is impossible to construct regular solids from hexagons, heptagons, octagons, nonagons, decagons, and other similar polygons."

At one place, Pacioli goes beyond Euclid. He presents what we have called the arithmetical proof for the incommensurability of the diagonal and side of the regular pentagon and credits Campanus with this proof:

XV. "Of its sixth unnameable elaboration. How no rational number can be so divided in this proportion that the parts are themselves rational."

In our section on beauty in mathematics we have seen that for the Greeks "symmetry," that is to say commensurability, was one of the most important characteristics of beauty. Seen that way, the EM ratio with its incommensurable parts is not at all beautiful for the Greek mind. Some German professor, however, thought otherwise. The term "golden section" (= goldener Schnitt) for the EM ratio seems to have originated in Germany in the first half of the nineteenth century. The German art historian A. Zeising [1855] was the first to "discover" the importance of the golden section for art history in his book *Aesthetik*. As far as I can see, his claims of almost magical effects of the golden section have never been taken seriously by his fellow art historians. Nevertheless, these efforts are popular with many people who delight in a mathematical explanation of beauty.

30.3 Philosophy

The regular solids are very often called the "Platonic solids" because of the prominent role they play in Plato's dialogue *Timaeus*. We have already quoted Plato's definition: "a solid that gives rise to a subdivision of the (circum–)sphere into equal and similar parts" (Tim. 55a). Plato associates the solids with the four elements and the cosmos as a whole. At the beginning of his discourse about the solids and the elements he says, "I believe that you will be able to follow me, for your education has made you familiar with the methods of science" (Tim 53c). This sentence is generally understood to be a hint to the relative novelty of the theory of the regular polyhedra in Plato's Academy. The dodecahedron is mentioned only at the end of the passage with one short sentence, which seems to be hard to translate: "There was yet a fifth combination which God used in the delineation of the universe with figures of animals" (Tim 55c). In his *Phaedo* 110b Plato puts the dodecahedron in a similar cosmological context. The dodecahedron excavated in Geneva with the names of the twelve signs of the zodiac inscribed on its faces (cf. above) seems to illustrate what Plato thinks about this solid.

The other four solids are associated with the classical elements: The tetrahedron (pyramid) with fire (pyr), the octahedron with air, the icosahedron with water, and the cube with earth. Plato continues to explain a sort of transition between the elements. Water, when heated, dissolves into 20 equilateral triangles that reassociate to become two octahedra and one tetrahedron, that is, in Plato's interpretation, air and heat, or in one word, steam. Crude as this may sound, it is historically the first attempt to describe natural phenomena by mathematical means and not by mythological explanations.

Plato has a curious way of composing the faces of the four solids from more elementary triangles. For the faces of the cube he combines four half–squares in order to get one square face of the cube. For the faces of the other solids he bisects equilateral triangles and combines six of the halves into a new equilateral triangle (Fig. 30.4).

That this step is not a minor issue can be inferred from the fact that Plato emphasizes its importance in a most unusual way: If anyone could claim, says he, that he had found a triangle "that is fairer for the construction of these bodies" (i.e., the regular solids corresponding to the four elements), "he, as friend rather than foe, is the victor" (Tim. 54a). The selection of the two kinds of elementary triangles as "the fairest" is, of course, inspired by the structure of

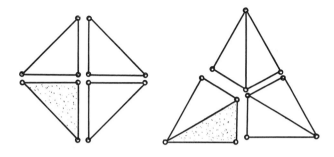

FIGURE 30.4

the regular solids, which they have to constitute. The solids are the prime candidates for aesthetic predication, *"for to no one will we concede that fairer bodies than these,* each distinct of its kind, *are anywhere to be seen.* Wherefore we must earnestly endeavor to frame together these four kinds of bodies which excel in beauty" (Tim. 53e–54a).

Strongly contrasting to the more cautious way Plato speaks about the fairness of the triangles, he is absolutely sure that there are no fairer bodies than the regular ones. As Plato says, there are many nonisosceles right triangles, and there is at first sight no obvious reason to single out the half of the equilateral triangle as the most beautiful one. The mathematical situation is quite different with respect to the solids. Plato has quoted the concept—presumably defined by Theaetetus—of regularity. Theaetetus is credited not only with the definition, but with the corresponding theorem as well: There are only five regular solids—the ones mentioned by Plato, including the dodecahedron. From the Pythagorean examples (see below) pyramid (tetrahedron), cube (hexahedron), and "the sphere of the twelve pentagons" (dodecahedron) Theaetetus proceeds—better, ascends—to the general concept and returns to the particulars, adding octahedron and icosahedron, and thus completes the list. This is a mathematical prototype of the dialecticians' procedure as stated in the *Republic* 511b. And it is a marvelous example of Plato's conception of form (idea) and participation: Each of the five solids participates in the idea of a regular solid, and conversely, the idea unfolds in exactly five particulars. Like the letters in *Philebus* 18c, it is impossible to know a single one without understanding them all together. Except for the dodecahedron, the regular solids are (mathematically) interesting as members of a family, not so much as individual polyhedra. For Plato the regular solids are the most beautiful ones because we can demonstrate by a priori reasoning that five and only five representations of the idea of a regular solid exist. We can show that the list of particulars of a certain idea is complete. No domain of empirical investigation can provide a similar case. Because of their philosophical significance the regular solids are the most beautiful ones.

30.4 Mathematics

We have already quoted Iamblichus on Hippasus of Metapontum (about 500–450 B.C.E.) about irrationality (Chapter 24). We repeat: "About Hippasus in particular they [the ancients] report that he belonged to the Pythagoreans, but because he was the first to make public the secret of the sphere of the twelve pentagons, he perished at sea. The fame of the discoverer, however, was his. . . ." (For the quotation see again Knorr [1975], p. 50 n. 6/7.) The "sphere of the twelve pentagons" is, of course, the dodecahedron. A scholion to Euclid gives us more information. It says that Euclid's Book XIII is about ". . . the five figures called Platonic, which, however, do not belong to Plato. Three of these five figures, the cube, pyramid, and dodecahedron, belong to the Pythagoreans, while the octahedron and icosahedron belong to Theaetetus." (See Heath, Introduction to Book XIII.)

The pyramid, the cube, and the "sphere of the twelve pentagons" have names from everyday language, the octahedron and the icosahedron have artificial mathematical names. From this and the appearance of dodecahedral pyrite crystals we conclude that the scholion is right and credit Hippasus with the first scientific study of the dodecahedron. Since he had to know, for its construction, the EM ratio of the diagonal and side of the regular pentagon, this story is in harmony with the other one about him as the discoverer of the phenomenon of incommensurability. At least he would have had a very good reason to try to find the EM ratio. Hippasus's construction of the dodecahedron may have been very similar to the one given by Euclid in Prop. XIII. 17.

The second stage in the mathematical history of the regular solids seems to have been what the anonymous scholiast to Euclid tells us about Theaetetus's discovery of the octahedron and the icosahedron. In accordance with a very detailed study by Eva Sachs [1917] and a recent one by William Waterhouse [1972], it is generally accepted that Theaetetus (1) defined the concept of regularity, (2) constructed the octahedron and the icosahedron, and (3) proved that there are only five regular solids. Theaetetus created the prototype of a mathematical theory; starting from a significant example, the dodecahedron, he proceeds to a general concept, and manages to classify all objects satisfying the definition. A detailed study of the objects, surviving in the second half of Euclid's Book XIII, completes the investigation.

The various definitions of regularity surviving from antiquity attest to the importance of the regular solids, certainly based for a good part on Plato's enthusiasm for them.

Interest in the regular solids was rekindled when knowledge of the writings of Plato and Euclid became more widespread during the late Middle Ages and the Italian Renaissance. Kepler (1571–1630) used them for his cosmological model, and Descartes (1596–1650) and Euler (1707–1783) studied polyhedra in general.

The modern part of the story of the regular solids begins roughly between 1800 and 1870, when crystallographers and mathematicians introduced the concept of a group into mathematics. Felix Klein, who was equally interested in group theory and complex function theory, wrote his famous treatise, *The Icosahedron and the Solution of Equations of the Fifth Degree*, in 1884. At the very beginning of this book he studies the symmetry groups of the regular solids in detail. These are finite subgroups of the special orthogonal group SO_3 of rotations of 3–space. Klein gives credit to Laguerre and Cayley for establishing the isomorphism of SO_3 with the group PSU_2 of homogeneous, linear, unitary transformations of determinant 1 of the

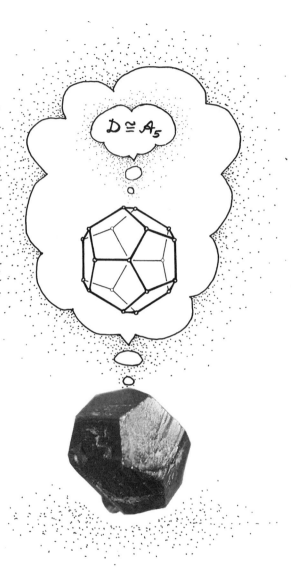

FIGURE 30.5

extended complex plane. This isomorphism connects geometry and function theory, thus giving rise to a new and very deep extension of the theory of regular polyhedra, which flourishes to this day. (See, for instance, Bättig and Knörrer [1991], Lamotke [1986], and Coxeter [1988])

Starting from geometrical figures, Klein finds the following finite subgroups (up to conjugacy and together with their respective subgroups) of SO_3:

- Dihedral groups as the symmetry groups of dihedrons, or double pyramids
- The tetrahedral group, which is isomorphic to the group $\mathcal{A}_4$ of even permutations of four symbols
- The octahedral group, which is the same as the group of the cube, isomorphic to the full group $\mathcal{S}_4$ of permutations of four symbols
- The icosahedral group (= dodecahedral group $\mathcal{D}$), which is isomorphic to the group $\mathcal{A}_5$ of even permutations of 5 symbols.

For the isomorphism between the dodecahedral group $\mathcal{D}$ and $\mathcal{A}_5$ (see Fig. 30.5), Klein uses the five cubes that can be inscribed in a regular dodecahedron as permuted symbols (as already observed by Hippasus and Euclid). By geometric arguments Klein proves the simplicity of the dodecahedral group $\mathcal{D}$. This proof is reproduced with minor modifications by Artmann [1988c].

In an earlier paper of 1875, Klein stated the completeness of the list of finite subgroups of SO_3 given above as a theorem. More generally, he considered finite subgroups of the groups of all isometries of Euclidean 3–space. He proved that such a group must be a subgroup of SO_3 and concluded "... and hence there are no other examples except the ones mentioned above." He furnished the proof for the subgroups of SO_3 in Chapter V, Section 2, of the book on the icosahedron [1884, pp. 128–130 of the English edition]. Nice elementary presentations of this proof are given by Coxeter [1988, pp. 70–72] and by Senechal [1990].

Klein's theorem puts the theory of regular solids in a totally new perspective; the abstract groups of symmetries suffice to recapture the polyhedra. The dihedral subgroups of SO_3 have an invariant subspace (the axis of the double pyramid), and all other finite subgroups of SO_3 are (conjugate to) subgroups of the symmetries of the tetrahedron, the octahedron, or the dodecahedron.

The Dodecahedron Again

The elementary building blocks of finite groups are the simple groups. These are groups without proper homomorphic images, or, equivalently, without proper normal subgroups. Groups of prime order p are simple, and these are the only abelian simple groups. The smallest nonabelian simple group has order 60; it is the symmetry group $\mathcal{D}$ of the dodecahedron (or icosahedron). As mentioned above, this group is isomorphic to the alternating group $\mathcal{A}_5$, and this is the first one in the series of simple groups $\mathcal{A}_n(n \geq 5)$. Several other series of finite simple groups are known, and moreover, there are some "sporadic" simple groups. In one of the most exciting stories in modern mathematics, in 1960–1980 group theorists completed the classification of all finite simple groups. (For more details see Conway [1980].) Just as at the beginning of Greek mathematics, the dodecahedron has again provided the first significant example for a great mathematical theory.

31

The Origin of Mathematics 16

The Origin of the *Elements*

Heiberg, the modern editor of the Greek original of the *Elements* and the most eminent Euclid scholar of his time, thought in 1904 that it was nearly impossible to reconstruct earlier versions of Greek mathematics from the *Elements* (Heiberg [1904], p. 4). Mathematical and stylistic analyses over the last 100 years have disproved Heiberg's claim and revealed a great deal about the prehistory of this monumental compendium of Greek mathematics. Our present, fairly complete, picture of the different contributions of pre–Euclidean mathematicians to the *Elements* is to a large extent due to the detailed studies begun by Becker in the 1930s, and continued by Neuenschwander on the geometrical books and by Mueller on the *Elements* as a whole. Heath and the other translators of the *Elements* provide valuable commentaries, but they are primarily concerned with individual definitions, theorems, and proofs. The global picture emerges from the investigation of the mathematical architecture

313

of the *Elements* in combination with the study of other ancient sources.

For the principal problems of such investigations we quote Mueller [1998, Ch. IV]:

> The *Elements* covers a wide variety of subjects. To a considerable extent, different subjects are treated in different books or sequences of books. One might say that the fundamental problem for determining the overall structure of the *Elements* is resolving the tension between the conception of the work as a collection of relatively independent treatises, lightly retouched by Euclid, and the conception of it as a relatively integrated systematic presentation of ideas and theories developed over two centuries. It is easy to opt for the second conception on the basis of a casual reading of the *Elements*, but once one begins a careful study of details, problems in that conception begin to emerge as discrepancies and logical gaps are noticed. If one then looks for discrepancies and gaps, one will find more; and if one adopts the first conception, one will find that there are even more features of the *Elements* which can be invoked to argue against its basic unity. Such hypercriticism is very valuable because it makes one more aware of the details of the text, but when it is conjoined with the attempt to find previously unknown aspects of pre–Euclidean mathematics it easily leads to conjectural fantasy. The position adopted here is that we will better understand the *Elements* as a scientific work if, while recognizing the genuine discrepancies and gaps in it, we work to understand the conceptual apparatus which made them less noticeable to Euclid and to virtually all pre–nineteenth–century readers than they are to us today. We know that Euclid was putting together the mathematical accomplishments of others; it is even possible that there is no mathematical result in the *Elements* ascribable to Euclid. But there is no reason to think that Euclid was not attempting to synthesize those results. The shortcomings of his synthesis should not lead us to turn our backs on the overall power of the Euclidean achievement.

We will leave it at that and, instead of many words, will use one picture (Fig. 31.1) to show the general architecture of the *Elements*.

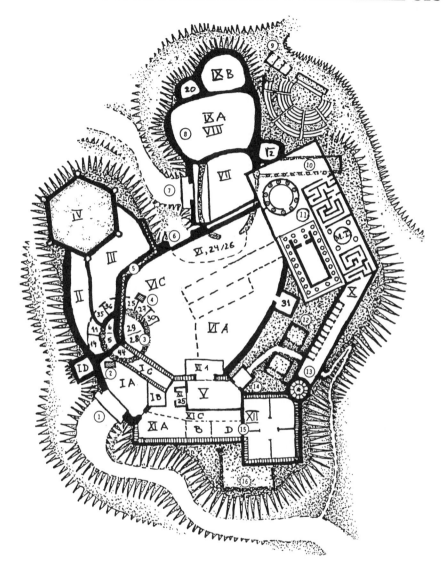

FIGURE 31.1 *Castrum Euclidis.* 1. Gate 1 to geometry. 2. Statue of Thales. 3. Museum of the Pythagoreans. 4. Statue of Pythagoras. 5. Murus contra proportiones sive adversus rationes. 6. Old gate(?). 7. New gate 2 to arithmetic. 8. Castle of Archytas. 9. Quarters of the musicians. 10. School of Theodorus. 11. Labyrinth and temple of Theaetetus. M.T. the Minotaurus. 12. Ruins of anthyphairesis. 13. the inexhaustible well. 14. Gate 3 to the general theory of proportions. 15. Palace of Eudoxus. 16. Ruins of "Conica"(?).

Notes

B. L. van der Waerden's book *Science Awakening* [1954] is the most important general introduction to ancient mathematics. Some of van. der Waerden's conclusions have been disputed, but on the whole his book is indispensible. Euclid's *Elements* have been analysed in every conceivable detail by Ian Mueller [1981]; his book has become fundamental for the study of Greek mathematics. David Fowler [1987/1998] discusses, among other things, the actual transmission of ancient mathematics on papyri and its relations to everyday calculation in Hellenistic Egypt. For the concept of proof, which I do not discuss specifically, compare the forthcoming book by R. Netz [1999].

Robin Hartshorne's *Companion to Euclid* [1997] is a course of geometry, based on Euclid's *Elements*, but is in its essence more oriented toward modern mathematics in a technical sense than towards history. The many very helpful comments by T. L. Heath in his translation of the *Elements* have already been mentioned in the "Notes to the Reader." Heath's two volumes on the history of Greek mathematics are equally important. A translation of the *Elements* into French by B. Vitrac brings Heath's comments up to date; its fourth and last volume is soon to appear.

Complete and up to date information on ancient mathematicians, e.g.; on Euclid and his works, can be found in the *Dictionary of Scientific Biography*.

1. General Historical Remarks

For the history and culture of ancient Greece in general, see Borbein [1995]. This book is being translated into several other languages.

3. The Origin of Mathematics 1

About Eudemus and Proclus in general see the introduction by Ian Mueller to Proclus–Morrow [1992], about Eudemus most recently, Vitrac [1996]. The etymology of the word "mathematics" is taken from Klein [1967], see also Proclus–Morrow pp. 37/38 about the origin of the word "mathematics."

4. Euclid Book I

The overwhelming majority of papers about the *Elements* do not treat anything beyond Prop. I. 4. Our emphasis is different. The use of Aristotle's remarks for an explanation of the proof of Prop. I. 5 is new. For detailed comments on Aristotle's remarks see Heath about Prop. I. 5.

Prop. I. 16 and its role in the composition of Book I have been studied by Artmann [1996]. For the work of Menelaos see Heath, *Greek Mathematics* II, 261–273.

About the consequences of Prop. I. 32 in philosophy see Schmitz [1997]. J. Steiner's proof of the Euler formula is reproduced in Artmann: *Lineare Algebra* pp. 320/21 (Basel: Birkhäuser 1991).

Prop. I. 47: Proclus (Proclus–Morrow p. 338) credits Euclid personally with this proof.

Adelbert von Chamisso (1781–1838):

Die Wahrheit

Die Wahrheit, sie besteht in Ewigkeit,
Wenn erst die blöde Welt ihr Licht erkannt:
Der Lehrsatz, nach Pythagoras benannt,
Gilt heute, wie er galt zu seiner Zeit.

Ein Opfer hat Pythagoras geweiht
Den Göttern, die den Lichtstrahl ihm gesandt;
Es taten kund, geschlachtet und verbrannt,
Ein Hundert Ochsen seine Dankbarkeit.

Die Ochsen seit dem Tage, wenn sie wittern,
Daß eine neue Wahrheit sich enthülle,
Erheben ein unmenschliches Gebrülle;

Pythagoras erfüllt sie mit Entsetzen;
Und machtlos, sich dem Licht zu widersetzen,
Verschließen sie die Augen und erzittern.

Delbrück's translation has been published by O. Taussky in the Mathematical Intelligencer 10 (1988) p. 53. For the quotation of Dodgson/Lewis Carroll, also from the *Intelligencer* (19 (1988), 31), I don't have an exact reference.

5. The Origin of Mathematics 2

For the pre-Euclidean theory of parallels see Dehn [1938], 15–18, and Thot [1966], 300. On Menaechmus see van der Waerden [1954]. Sources for the history of the parallel axiom are to be found in the book Becker [1954] for the time before 1860, and in Stillwell [1996] for papers from Beltrami and later ones. Stillwell's introductory notes to the sources are very helpful for a first understanding of hyperbolic geometry. Engel–Staeckel [1895] are very complete up to their time. Hartshorne Chap. 2 discusses Pasch's and Hilbert's axiom systems in detail.

6. The Origin of Mathematics 3

The most important books about Pythagoras and the Pythagoreans are Burkert [1972] and van der Waerden [1979]. Information about the Greek architects comes from Gruben [1984] and Knell [1988], for Eupalinus see again van der Waerden [1954]. The pentagram as a medical symbol is studied by Schouten [1968].

For the coins see Artmann [1990].

7. Euclid Book II

Heath, van der Waerden, and Neuenschwander assume a Pythagorean origin of Book II, but Fowler [1987] thinks otherwise. In purely mathematical terms, one is tempted to propose the following sequence of events:

- The division of a line in "extreme and mean ratio" was found by Pythagoreans as in Prop. XIII. 8 and was constructed via the application of areas. (See "Second analysis and synthesis of the pentagon" in our discussion of Book IV.)
- The application of areas was replaced by the methods in the first part of Book XIII, hiding the original motivation.
- The methods from the first part of Book XIII were generalized, streamlined, and supplemented with other material, resulting in Book II.

(See Fowler [1987/1998] for a very different point of view.)

8. The Origin of Mathematics 4

The ancient history of this problem has been worked out in every conceivable detail by Knorr [1986], from whom most of the quotations are taken (esp. pp. 25 ff.).

Dante: *The Divine Comedy. Paradise*, Canto XXXIII lines 133–136. The Italian original is

Qual è il geomètra che tutto s'affige
per misurar lo cherchio, e non ritrova,
pensando, quel principio ond'elli indige,
tal era io a quella vista nova ...

For the further history of π, see M. Kline [1972], 593, or the many original documents in Berggren et al. [1997].

9. Euclid Book III

Mueller [1981], 183, points to the limited foundations of many of the propositions in parts A and B of Book III. On Pages 195–202 the notions of equality for circles and for the similarity of segments are analyzed in depth. Part C_2, about chords and angles, is obviously preparatory for Book IV.

10. The Origin of Mathematics 5

For a proper understanding of a mathematical theory one always has to consider two directions of thought: from the whole to the examples, and back from the examples to the whole. But in general, the whole will be more than the sum of its parts!

For the role of the cross ratio in hyperbolic geometry see, for instance, Jones, G. A. and D. Singerman: *Complex Functions*. Cambridge U. Press 1987.

11. Euclid Book IV

The construction of the incircle: As in this case, Euclid often takes a definite example but constructs the proof in such a way that the general procedure becomes obvious. See, for example, Props.II.1, V.12, VI.20, VII.33, and elsewhere.

About the use of a marked ruler cf. Heath's notes to Prop.II.6; Pappus–Jones Book VII.27/28 (Jones I, p. 112); and for an algebraic analysis Hartshorne [1997] sections 30/31. About the restriction to ruler and compass see Steele [1936].

13. The Origin of Mathematics 7

For Gauss and the 17–gon see Dunnington [1955], esp. p. 28. A modern construction of the 17-gon is carried out by Hartshorne (section 29).

14. Euclid Book V

For the Pythagorean theory of harmonics and music see Burkert [1972], Ch. V and van der Waerden [1954]. For architecture cf. Gruben [1984], Knell [1988], and Heisel [1993]. The Canon of Polyklet is studied by v. Steuben [1973]. Much has been written about proportions in music, but very little about possible relations between architecture and geometry. Burkert, Ch. V, points to the origin of mathematical terminology from the building trade (like "orthogonal"). It seems likely that architecture was more important for geometry than field measurement, even in Egypt.

About proportions in a modern version: For the configuration theorem of Pappus as the strongest geometrical axiom (in projective planes) see Pickert [1975].

15. Euclid Book VI

The application of areas: As presented by Euclid, Props.VI.28/29 and their proofs have taken such a baroque appearance as to obscure what is really going on. One cannot understand what the generalization to parallelograms should be good for. Moreover, the decisive step of constructing a mean proportional (i.e., square root) is hidden in Prop. VI.25. About Prop. VI.25: For the (more general) role of water in Aristotle's natural philosophy see Böhme [1998].

16. The Origin of Mathematics 8

For a more detailed discussion of the generalization from rectangles to parallelograms see Artmann [1988 a], 129–131. See also the note above about the application of areas.

17. Euclid Book VII

The historical background: Here are the lines by Aeschylus in their context. Prometheus speaks:

> But hearken to the miseries that beset mankind—how that they were witless erst and I made them to have sense and be endowed with reason. Nor will I speak to upbraid mankind, but to set forth the friendly purpose that inspired my boons.
>
> First of all, though they had eyes to see, they saw to no avail; they had ears, but understood not; but, like to shapes in dreams, throughout their length of days, without purpose they wrought all things in confusion. Knowledge had they neither of houses built of bricks and turned to face the sun, nor yet of work in wood; but dwelt beneath the ground like swarming ants, in sunless caves. They had no sign either of winter or of flowery spring or of fruitful summer, whereon they could depend, but in everything they wrought without judgment, until such time as I taught them to discern the risings of the stars and their settings, ere this ill distinguishable.
>
> *Aye, and numbers, too, chiefest of sciences, I invented for them,* and the combining of letters, creative mother of the Muses' arts, wherewith to hold all things in memory. I, too, first brought brute beasts beneath the yoke to be subject to the collar and the pack–saddle, that they might bear in men's stead their heaviest burdens;

and to the chariot I harnessed horses and made them obedient to
the rein, to be an adornment of wealth and luxury. 'Twas I and no
one else that contrived the mariner's flaxen-winged car to ream
the sea.

Wretched that I am—such are the inventions I devised for
mankind, yet have myself no cunning wherewith to rid me of
my present suffering.

(Aeschylus: *Prometheus Bound*, lines 440–471)

Money: The date of the first coins from Aigina (an island near
Athens) is about 580 B.C.E.

Detailed studies of the arithmetical books are, as always, van
der Waerden [1954] (and many more papers), Mueller [1981], Itard
[1961], and Taisbak [1971], to whom I am indebted very much. For
more about mythological and enlightened historical texts see Vitrac
[1996].

Proportion for numbers: In my interpretation of VII.4 I am following Zeuthen as reported by Mueller [1981], p. 62. The role of
reduced fractions (*minimi*) is discussed in detail by Taisbak [1971].

Propositions about proportions: About the different definitions
of proportion for (geometric) magnitudes and numbers see Mueller
[1996].

Proportions and products: For the use of square grids and similar
auxiliary means in the arts see Müller [1973].

The greatest common divisor. It is not hard to prove the uniqueness of *minimi* directly starting from the Euclidean algorithm *EA*.
Assume $r : s = t : u$ and $gcd(r, s) = 1 = gcd(t, u)$. From *EA*
one gets integers x, y such that $xr + ys = 1$. This can be used
to show $t = r(xt + yu)$; hence r divides t, and so on, resulting in
$r = t$ and $s = u$. The existence of minimi could then be secured by
Prop. VII. 33; but it seems doubtful that this would be the right interpretation of Euclid's theory. Note that Euclid does not care about
the uniqueness of the *gcd* of two numbers either. For the use of induction in Greek mathematics see Fowler [1994] and Unguru [1991,
1994].

21. Euclid Book IX

Odd and even numbers. Plato speaks about "the odd and the even" in *Charmides* 166 a 3–11, *Republic* 510 c 2–6, *Theaetetus* 198 a 5–9. Waschkies [1989], 276/277, follows Burkert [1972], Ch. VI, in calling this part of Book IX a piece of undergraduate work from Plato's Academy. (I agree.)

22. The Origin of Mathematics 11

Besides Proclus, Hardy, von Neumann, and Rota, many other mathematicians have expressed similar opinions. We have quoted Plotinus on beauty in our section on "Polygons after Euclid" and emphasized arguments that are characterized by clarity, conciseness, and a certain surprise at several other occasions. Here are some further papers of interest to the topic: Borel [1981], Hasse [1998], Knopp [1985], Krull [1987], and Manin [1998].

23. Euclid Book X

About the difficulties concerning the two different definitions of proportion see again Mueller [1996].

24. The Origin of Mathematics 12

In the words of one reviewer, much ink has been spilled about incommensurability in Greek mathematics. We follow suit. The reader should always keep in mind what is said in the summary about fact and fiction.

Geometrical proofs: the square. Side and diagonal numbers for the square are discussed by Knorr [1975], 29–36.

The pentagon: For the pentagon/pentagram as a medical symbol see Schouten [1968]. Vitruvius in his "De Architectura" points out that Epidaurus is the only Greek theater with a pentagonal symmetry. See also Gruben [1984] and Knell [1988].

25. Euclid Book XI

The duplication of the cube: For the extended literature on this famous problem, see Knorr [1986]. See also the remarks made about the plane analogue following Prop. VI. 25.

26. The Origin of Mathematics 13

The definition of a sphere: There is a relatively old fragment of the philosopher Parmenides (shortly after 500 B.C.E.) describing a sphere: "a well rounded sphere, from a center in all directions equally extended" (Parmenides Diels–Kranz [1956] 28 B 8). The German original of the quotation of Musil is: "Und so will jedes Wort wörtlich genommen werden, sonst verwest es zur Lüge, aber man darf keines wörtlich nehmen, sonst wird die Welt ein Tollhaus." (*Der Mann ohne Eigenschaften* [1981], p. 749.)

27. Euclid Book XII

The pyramid: For a comparative study of modern formulas and Euclid's way of expression, this subsection of the *Elements* may be particularly well suited. In principle, Prop. XII. 8 would enable Euclid to generalize the result to convex solids by suitable subdivisions as was done for the plane case in Prop. VI. 26, but that proof was already complicated enough.

29. Euclid Book XIII

Some guesses about part A as a possible forerunner of Book II are in the notes for Book II. The icosagon at Delphi: Bousquet [1993] is overly optimistic about the role of geometry in the construction of this temple. Besides the obvious number 20 he find ratios like $\sqrt{\pi} : \sqrt{2}$ for parts of the building, which has nothing to do with Greek mathematics. However, Bousquet was the first archeologist to point out the Fibonacci numbers in the theater of Epidauros. The number of columns in the inner circle is something of a mystery. The plan shows an 11-gon, counting one "invisible" column at the place of the door. There is a similar problem with the number of columns in the Tholos in Epidaurus (Knell [1988], 228/29). It has 26 (two times the Fibonacci number 13) on the outside and 14 in the interior cella. The Tholos in Epidaurus was built some 30 years later than the Tholos in Delphi.

The construction of the dodecahedron. Note that along with the dodecahedron a regular pentagon is constructed depending only on the earlier propositions in Book XIII. No use is made of the construction of the pentagon in Book IV.

30. The Origin of Mathematics 15

Art: A few reproductions of Leonardo's and Jamnitzer's pictures can be found in Toepell [1991]. Philosophy: See Artmann–Schaefer [1993] for more on Plato's "fairest triangles". Mathematics: Toepell [1991] has a rather complete list of literature for the more elementary parts of the history. The dream of abstraction has already been dreamed by a cube in Saul Steinberg's "Labyrinth".

31. The Origin of Mathematics 16

For more details see Artmann [1991]. An earlier version of the "Castrum Euclidis" has appeared in Artmann [1988 a].

Bibliography

Aristotle 1991. *The Complete Works of Aristotle.* J. Barnes, ed. Bollingen series LXXI 1/2 Princeton University Press.

Artmann, B. 1984. "Hippasos und das Dodekaeder." *Mitteilungen des Mathematischen Seminars der Universität Giessen* **165**, 103–121.

Artmann, B. 1985. "Über voreuklidische Elemente, deren Autor Proportionen vermied." *Archive for History of Exact Sciences* **33**, 291–306.

Artmann, B. 1988. *The Concept of Number.* Chichester: Ellis Horwood/John Wiley.

Artmann, B. 1988a. "Über voreuklidische Elemente der Raumgeometrie aus der Schule des Eudoxos." *Archive for History of Exact Sciences* **39**, 121–135.

Artmann, B. 1988b. "Die Mathematik in der Akademie Platons. Über eine neue Rekonstruktion von D.H. Fowler." *Mathematische Semesterberichte* **35**, Heft 2, 162–182.

Artmann, B. 1988c. "A simple proof for the simplicity of $\mathcal{A}_5$," *The American Mathematical Monthly,* Vol 95, n. 4, 344–349.

Artmann, B. 1990. "Mathematical motifs on Greek coins." *The Mathematical Intelligencer* **12**, 43–50.

Artmann, B. 1991. Euclid's Elements and its Prehistory. *In Peri Ton Mathematon,* Mueller, I. ed. *Apeiron* 24 (4), 1–47.

329

Artmann, B. and Schäfer, L. 1993. On Plato's "Fairest Triangles" (Tim. 54a). *Historia Mathematica* **20**, 255–264.

Artmann, B. 1993. Roman Dodecahedra. *Math. Intelligencer* **15**, no. 2, 52–53.

Artmann, B. 1994a. A proof for Theodorus' theorem by drawing diagrams. *Journal of Geometry* **49**, 3–35.

Artmann, B. 1994b. Symmetry Through the Ages. Highlights from the History of Regular Polyhedra. In: J. M. Anthony, ed.: *In Eves Circles*. 139–148. MAA Notes No 34, Washington, D.C.

Artmann, B. 1996. *Euclid's proposition Elements I.16 and its consequences for the genesis of Elements Book I.* In: T. Berggren, ed.: Proceedings of the third international conference on ancient mathematics in Delphi, Greece. Simon Fraser University, 3–12.

Artmann, B. 1996. A Roman Icosahedron Discovered. *Amer. Math. Monthly* **103**, no 2, 132–133.

Artmann, B. and I. Mueller 1997. Plato and Mathematics. *Mathematische Semesterberichte* **44**, 1–17.

Bättig, D. and Knörrer, H. 1991. *Singularitäten*, Basel: Birkhäuser.

Becker, O. 1933. "Eudoxos–Studien I. Eine voreudoxische Proportionenlehre und ihre Spuren bei Aristoteles und Euklid." *Quellen und Studien zur Geschichte der Mathematik, Astronomie und Physik* Abteilung B **2**, 369–387.

Becker, O. 1934. "Die Lehre vom Geraden und Ungeraden im neunten Buch der euklidischen Elemente." *Quellen und Studien zur Geschichte der Mathematik, Astronomie und Physik* Abteilung B **3**, 533–553.

Becker, O. 1954. *Grundlagen der Mathematik in geschichtlicher Entwicklung.* Freiburg: Alber.

Beckmann, F. 1967. "Neue Gesichtspunkte zum 5. Buch Euklids." *Archive for History of Exact Sciences* **4**, 1–144.

Berggren, J. L. 1984. "History of Greek mathematics: a survey of recent research." *Historia Mathematica* **11**, 394–410.

Berggren, J. L. and R. S. D. Thomas 1996. *Euclid's Phaenomena.* New York: Garland.

Berggren, J. L., J. Borwein, P. Borwein 1997. *Pi: A Source Book.* New York: Springer.

Böhme, G. 1998. Kontinuität als Phänomen in der Philosophie des Aristoteles. In: J. Beaufort and P. Prechtl, eds.: *Rationalität und Prärationalität.* p. 37–45. Königshausen und Neumann.

Borbein, A. H. (ed.) 1995. *Das alte Griechenland.* München: Bertelsmann.

Borel, A. 1983 *Mathematics: Art and Science.* Carl Friedrich von Siemens Stiftung. Math. Intelligencer 5(4) 1983, 9–17.

Bousquet, J. 1993. La tholos de Delphes et les mathématiques préeuclidiennes. *Bull de correspondance Hellénique* 107, 283–313.

Bretschneider, C. A. 1870. *Geometrie und die Geometer vor Euklides.* Leipzig: Teubner (Reprint Wiesbaden: Sändig 1968).

Burkert, W. 1972. *Lore and Science in Ancient Pythagoreanism,* translated by Edwin L. Minar Jr. Cambridge, Mass.: Harvard University Press. (Original German edition: *Weisheit und Wissenschaft.* Nürnberg: Hans Carl 1962)

Burnyeat, M. F. 1978. "The philosophical sense of Theaetetus' mathematics." *Isis* **69**, 489–513.

Burnyeat, M. F. 1987. Platonism and Mathematics. A Prelude to Discussion. In: A. Graeser, ed.: *Mathematics and Metaphysics in Aritotle.* Bern: Haupt, 213–240.

Cervi–Brunier, I. 1985. "Le dodécaèdre en argent trouveé à Saint Pierre de Genève," *Zeitschrift für Schweizerische Archäologie und Kunstgeschichte,* (Band 42, Heft 3) 153–156.

Charbonneauy, J. and G. Gottlob 1925. *La tholos. Fouilles de Delphes, tome II.* Le sanctuaire d'Athèna. Paris: E. de Broccard.

Cherniss, H. 1951. Plato as Mathematician. *The Review of Metaphysics,* Vol. IV, No. 3, 395–425.

Conway, J. H. 1980. "Monsters and Moonshine," *The Mathematical Intelligencer 2,* no. 4, 165–171.

Cornford, F. M. 1937. *Plato's Cosmology.* London: Routledge.

Cornford, F. M. 1965. Mathematics and dialectic in Republic VI–VII. In: R. E. Allen, ed.: *Studies in Plato's Metaphysics.* London: Routledge and Kegan Paul, 61–96.

Coxeter, H. S. M. 1988. "Regular and Semi–regular Polyhedra." In M. Senechal and G. Fleck, eds., *Shaping Space, A Polyhedral Approach,* Birkhäuser: Basel.

Critchlow, K. 1979. *Time Stands Still,* London: Gordon Fraser.

Davis, Ph. J. and R. Hersh 1981. *The Mathematical Experience.* Boston: Birkhäuser.

Dehn, M. 1938. Beziehungen zwischen der Philosophie und der Grundlegung der Mathematik im Altertum. *Quellen und Studien zur Geschichte der Mathematik, Astronomie und Physik,* Abt. B, Band 4 (1938) 1–28.

Deonna, W. 1954. Les dodécaèdres gallo–romains en bronze, ajourés et bouletés, *Bull. Assoc. Pro Aventica* 16, 19–89.

Dictionary of Scientific Biography. New York: Scribner's. 1970–1980.

Diels, H. 1956. (Diels–Kranz) *Die Fragmente der Vorsokratiker.* 6th ed., 3 vols. Herausgegeben von Walther Kranz. Berlin: Weidmann.

Dunnigton, G. W. 1995. *Carl Friedrich Gauss. Titan of Science.* New York: Hafner.

Ebert, M. *Reallexikon der Vorgeschichte.* 3 Vols.

Ebert, T. 1974. *Meinung und Wissen in der Philosophie Platons,* Berlin: de Gruyter.

Euclid–Heath, T. L. 1926. *The Thirteen Books of Euclid's Elements.* 2nd ed., 3 vols. Cambridge, England: Cambridge University Press.

Euclid (Data) Thaer, C. 1962. *Die* Data *von Euklid.* Übersetzt und herausgegeben von C. Thaer. Berlin: Springer.

Euclid (Elements) Thaer, C. 1969. *Die Elemente.* Herausgegeben und ins Deutsche übersetzt von Clemens Thaer. Darmstadt: Wissenschaftliche Buchgesellschaft.

Euclid–Vitrac, B. 1990, 1994. *Euclide: Les Éléments.* Presses Universitaires de France: Paris. (Vol I 1990, Vol II 1994, Vol III 1998, Vol IV, to appear).

Engel, F. and P. Staeckel 1895. Die Theorie der Parallellinien von Euklid bis auf Gauss. Leipzig, Teubner.

Feldhag, R. and S. Unguru. *Greek mathematical discourse: some examples of tensions and gaps.* In: T. Berggren, ed.: Proceedings of the third international conference on ancient mathematics in Delphi, Greece. Simon Fraser University.

Fowler, D. H. 1980. "Book II of Euclid's Elements and a pre-Eudoxan theory of ratio." *Archive for History of Exact Sciences* **22**, 5–36.

Fowler, D. H. 1987. *The Mathematics of Plato's Academy. A New Reconstruction.* Oxford: Oxford University Press. (Second revised edition to appear in 1999).

Fowler, D. H. 1992. An invitation to read Book X of Euclid's Elements. *Historia Mathematica* **19**, 233–264.

Fowler, D. H. 1994. Could the Greeks have used mathematical induction? Did they use it? *Physis* **31**, 253–265.

Frajese, A. 1963. *Platone e la matematica nel mondo antico* (Universale Studium testi e documenti 4). Roma: Editrice studium.

Freudenthal, H. and van der Waerden, B. L. 1947. "Over een bewering van Euclides." *Simon Stevin* **25**, 115–121.

Friedländer, P. 1960. *Platon.* Band III: Die Platonischen Schriften. Berlin: De Gruyter.

Gauss, C. F. 1801. *Disquisitiones arithmeticae.* (German translation: Untersuchungen über höhere Arithmetik. Berlin: Springer 1889)

Grattan–Guinnes, I. 1996. "Numbers, Ratios, and Proportions in Euclid's Elements: How Did He Handle Them?" *Historia Mathematica* 23, 355–375.

Gruben, G. 1984. *Die Tempel der Griechen.* Darmstadt: Wissenschaftliche Buchgesellschaft.

Hardy, G.H. and E.M. Wright 1954. *An introduction to the theory of numbers.* 3rd. ed. London, Oxford Univ. Press.

Hardy, G.H. 1967. *A Mathematician's Apology.* Cambridge University Press. (Originally 1940).

Hartshorne, R. 1997. *Companion to Euclid.* A course of geometry, based on Euclid's Elements and its modern descendants. Berkeley Math. Lecture Notes, vol 9. Providence, R.I., American Mathematical Society.

Haselberger, L. 1985. *The Construction Plans for the Temple of Apollo at Didyma.* Scientific American Dec, 1985, 114–122.

Hasse, H. 1997. Mathematik als Wissenschaft, Kunst und Macht. *Mitt. Deutsche Mathematiker Vereinigung* (4/97) 28–38.

Heath, T. L. 1910. *Diophantus of Alexandria.* Cambridge University Press.

Heath, T. L. 1921. *A History of Greek Mathematics.* 2 vols. Oxford: Clarendon.

Heath, T. L. 1949. *Mathematics in Aristotle.* Oxford: Clarendon.

Heiberg, J. L. 1904 "Mathematisches zu Aristoteles" *Abhandlungen zur Geschichte der mathematischen Wissenschaften* **18**, 1–49.

Heisel, J.P. 1993. *Antike Bauzeichnungen.* Darmstadt: Wissenschaftliche Buchgesellschaft.

Hilbert, D. 1899. *Grundlagen der Geometrie.* 10th ed. Stuttgart: Teubner 1968.

Hilbert, D. 1899 (1902). *Foundations of Geometry.* Chicago: Open Court 1902. First German edition: *Grundlagen der Geometrie* 1899 Leipzig: Teubner. Many subsequent editions.

Hofmann, J. E. 1926. "Ein Beitrag zur Einschiebungslehre." *Zeitschrift für mathematischen und naturwissenschaftlichen Unterricht* **57**, 433–442.

Høyrup, J. 1989. "Zur Frühgeschichte algebraischer Denkweisen." *Mathematische Semesterberichte* **36**, 1–46.

Itard, J. 1961. *Les livres arithmétiques d'Euclide.* Paris: Hermann.

Jacobs, H.R. 1987. *Geometry.* 2nd ed. New York: Freemann and Co.

Jamnitzer, W. 1568/1973. *Perspectiva corporum regularium.* Reprint Akademische Verlagsanstalt, Graz 1973.

Käppel, L. 1989. "Das Theater von Epidaurus." *Jahrbuch des deutschen archäologischen Instituts* **104**, 83–106.

Klein, E. 1967. *A Comprehensive Etymological Dictionary of the English Language,* 2 vols. Elsevier.

Klein, F. 1884. *Vorlesungen über das Ikosaeder und die Auflösung der Gleichungen vom fünften Grade,* Leipzig: Teubner, English translation, New York: Dover (1956).

Kline, M. 1972. *Mathematical Thought from Ancient to Modern Times.* New York: Oxford University Press.

Knell, H. 1988. *Architektur der Griechen.* 2nd ed. Darmstadt: Wissenschaftliche Buchgesellschaft.

Knopp, K. 1985. Mathematics as a cultural activity. *Math. Intelligencer* **7** (1) 7–14, 21.

Knorr, W. 1975. *The Evolution of the Euclidean* Elements. Dordrecht and Boston: Reidel.

Knorr, W. 1985. "Euclid's tenth Book: an analytic survey." *Historia Scientiae* **29**, 17–35.

Knorr, W. 1986. *The Ancient Tradition of Geometric Problems.* Boston, Basel and Stuttgart: Birkhäuser.

Krull, W. 1987. The aesthetic viewpoint in mathematics. *Math. Intelligencer* **9** (1) 48–52.

Lamotke, K. 1986. *Regular Solids and Isolated Singularities*, Wiesbaden: Vieweg.

Lasserre, F. 1990. *La naissance des mathématiques à l'èpoque de Platon* (Vestigia 7). Fribourg, Suisse: Editions universitaires.

Lindemann, F. 1896. Zur Geschichte der Polyeder und der Zahlzeichen. *Sitzungsber. der math.-phys. Klasse der Kgl. baierischen Akad. d. Wiss. XXVI*, 625–783.

Maass, M. 1993. *Das antike Delphi.* Darmstadt: Wissenschaftliche Buchgesellschaft.

Malmendier, N. 1975. "Eine Axiomatik zum 7. Buch der *Elemente* von Euklid." *Mathematisch-Physikalische Semesterberichte* **22**, 240–254.

Manin, Y. 1998. Interview: *Mitt. Deutsche Mathematiker Vereinigung* (2/98) 40–44.

Mehl, A. 1997. Medien im antiken Griechenland—Zur Bilder-Welt des Kindes. In: M. Liedtke, ed., *Kind und Medien*, 31–60. Bad Heilbrunn: J. Klinkhardt.

Mittelstrass, J. 1985. *Die geometrischen Wurzeln der platonischen Ideenlehre*, Gymnasium 92, 399–418.

Moise, E. E. 1974. *Elementary geometry from an advanced standpoint.* 2nd. ed. Reading: Addison-Wesley.

Mueller, I. 1981. *Philosophy of Mathematics and Deductive Structure in Euclid's* Elements. Cambridge, Mass.: MIT Press.

Mueller, I. 1991a. Mathematics and Education: Some Notes on the Platonic Program. In: I. Mueller, ed: *Peri Ton Mathematon*, Apeiron Vol. XXIV No. 4, 85–104.

Mueller, I. 1991b. On the notion of a mathematical starting point in Plato, Aristotle, and Euclid: In: A. C. Bowen, ed.: *Science and philosophy in classical Greece.* New York: Garland, 59–97.

Mueller, I. 1992. Mathematical method and philosophical truth. In: Richard Kraut, ed.: *The Cambridge Companion to Plato.* Cambridge, England: Cambridge University Press, 170–199.

Mueller, I. 1996. *Euclid as a blundering schoolmaster: a problem of proportion.* In: T. Berggren, ed.: Proceedings of the third international conference on ancient mathematics in Delphi, Greece. Simon Fraser University, 145–152.

Mueller, I. 1997. Greek arithmetic, geometry and harmonics: Thales to Plato. *Routledge History of Philosophy*, Vol. I. Edited by C. C. W. Taylor. p. 271–322. London and New York: Routledge.

Mueller, I. 1998. "Euclid's Elements from a philosophical perspective." To appear in: Mexican encyclopedia of the history and philosophy of mathematics.

Müller, H. W. 1973. Der Kanon in der ägyptischen Kunst. In: *Der "vermessene" Mensch*. München: Heinz Moos Verlag.

Musil, R. 1981. *Der Mann ohne Eigenschaften.* (The Man Without Qualities) Reinbek: Rowohlt.

Netz, R. 1999. (in press) *The Shaping of Deducation in Greek Mathematics.* Cambridge Univ. Press.

Neuenschwander, E. 1972. "Die ersten vier Bücher der *Elemente* Euklids." *Archive for History of Exact Sciences* **9**, 325–380.

Neuenschwander, E. 1973. "Beiträge zur Frühgeschichte der griechischen Geometrie." *Archive for History of Exact Sciences* **11**, 127–133.

Neuenschwander, E. 1975. "Die stereometrischen Bücher der *Elemente* Euklids." *Archive for History of Exact Sciences* **14**, 91–125.

Nouwen, R. 1993. *De Gallo-Romeinse Pentagon-Dodecaeder:* Mythe en Enigma. Publications of the Gallo-Romains Museum Tongeren (Belgium) Nr. 45, Hasselt.

Pacioli, Fra Luca 1509. *Divina Proportione*, Venedig. Neuherausgegeben, übersetzt und erläutert von Constantin Winterberg, Wien 1889.

Pappus–Jones 1986. *Pappus of Alexandria. Book 7 of the Collection.* Edited with translation and Commentary by A. Jones. New York: Springer.

Petronotis, A. 1969. *Bauritzlinien und andere Aufschnürungen am Unterbau griechischer Bauwerke in der Archaik und Klassik.* Dissertation TH München.

Pickert, G. 1975. *Projektive Ebenen.* 2nd. Springer: Berlin Heidelberg New York.

Plato 1989. *The collected dialogues.* E. Hamilton and H. Cairns, ed. Bollingen series LXXI, Princeton University Press.

Plotinus 1952. *The Six Enneads.* Translated by Stephen MacKenna and B.S. Page. Great Books of the Western World 17. The University of Chicago.

Proclus–Morrow, G. R. 1992. *A Commentary on the First Book of Eulcid's* Elements. 2nd. ed. Translated and edited by G. R. Morrow. Princeton: Princeton University Press.

Proust, M. 1981. *Remembrance of Things Past.* Vol III. (The Captive, The Fugitive, Time Regained) Translated by: C. K. Scott Moncrieff, Terence Kilmartin, Andreas Mayor. London: Chatto & Windus (French original 1927).

Ptolemaius–Kunitzsch, P. 1974. *Der Almagest: Die* Syntaxis Mathematica *des Claudius Ptolemäus in arabisch–lateinischer Überlieferung.* Wiesbaden: Harrassowitz.

Ritchi, Graham and Anna 1981. *Scotland* (Archeology), London.

Rota, J.C. 1997. *Indiscrete Thoughts.* Basel: Birkhäuser.

Sachs, E. 1917. *Die fünf platonischen Körper.* Philologische Untersuchungen **24**. Berlin: Weidmannsche Buchhandlung.

Schmitz, M. 1997. *Euklids Geometrie und ihre mathematiktheoretische Grundlegung in der neuplatonischen Philosophie des Proklos.* Würzburg: Königshausen & Neumann.

Schouten, J. 1968. *The pentagram as a medical symbol.* Niewkoop: De Graaf.

Schreiber, P. 1987. *Euklid.* Leipzig, Teubner.

Senechal, M. 1990. "Finding the finite groups of symmetries of the sphere," *Amer. Math. Monthly,* vol. 97, 329–335.

Steele, A. D. 1936. "Über die Rolle von Zirkel und Lineal in der griechischen Mathematik." *Quellen und Studien zur Geschichte der Mathematik, Astronomie und Physik* Abteilung B **3**, 288–369.

Steuben, H. v. 1973. *Der Kanon des Polyklet.* Tübingen: Wasmuth.

Stillwell, J. 1996. *Sources of Hyperbolic Geometry.* Providence, R.I., Amer. Math. Soc.

Stillwell, J. 1998. *Numbers and Geometry.* New York: Springer.

Taisbak, C. M. 1971. *Division and Logos: The Arithmetical Books of the* Elements. Odense: Odense University Press.

Taisbak, C. M. 1976. "Perfect numbers. A mathematical pun?" *Centaurus* **20**, 269–275.

Taisbak, C. M. 1982. *Coloured Quadrangles: A Guide to the Tenth Book of Euclid's* Elements. Copenhagen: Museum Tusculanum Press.

Thompson, F. H. 1970. Dodecahedrons again, *The Antiquaries Journal* 2, 93–96.

Toepell, M. 1991. Platonische Körper *Der Mathematikunterricht 37*, Heft 4, 45–79.

Toth, I. 1967. "Das Parallelenproblem im Corpus Aristotelicum." *Archive for History of Exact Sciences* **3**, 249–422.

Tymoczko, T. 1986. Making room for Mathematicians in the Philosophy of Mathematics. *Math. Intelligencer* Vol. 8, No. 3, 44–50.

Unguru, S. 1975. "On the need to rewrite the history of Greek mathematics." *Archive for History of Exact Sciences* **15**, 67–114.

Unguru, S. and Rowe, D. E. 1981, 1982. "Does the quadratic equation have Greek roots? A study of geometric algebra, application of areas, and related problems." *Libertas Mathematica* **1**, 1–49 and **2**, 1–62.

Unguru, S. 1991. Greek mathematics and Greek induction. *Physis* **28**, 273–289.

Unguru, S. 1994. Fowling after induction. *Physis* **31**, 267–272.

van der Waerden, B. L. 1947. "Die Arithmetik der Pythagoreer." *Mathematische Annalen* **120**, 127–153, 676–700.

van der Waerden, B. L. 1954. *Science Awakening*, translated by Arnold Dresden. Groningen: Noordhoff.

van der Waerden, B. L. 1978. "Die Postulate und Konstruktionen in der frühgriechischen Geometrie." *Archive for History of Exact Sciences* **18**, 343–357.

van der Waerden, B. L. 1979. *Die Pythagoreer*. Zürich: Artemis.

Vitrac, B. 1996. *Mythes (et réalités?) dans l'histoire des mathématiques grecques anciennes.* In: Goldstein, Catherine (ed.) al. L'Europe mathématique: histoires, mythes, identités. Issu d'un colloque-satellite du Congrés Européen de Mathématiques, Paris, France du 3 au 6 avril 1992. Paris: Éditions de la Maison des Sciences de l'Homme, 31–51.

Vlastos, G. 1991. *Socrates, Ironist and Moral Philosopher* (Cornell Studies in Classical Philology 50), Ithaca, New York: Cornell University Press.

von Fritz, K. 1945. "The discovery of incommensurability by Hippasos of Metapontum." *Annals of Mathematics* **46**, pp. 242–264.

von Fritz, K. 1971. *Grundprobleme der Geschichte der antiken Wissenschaft*. Berlin: de Gruyter.

v. Gerkan–Müller Wiener 1961. *Das Theater von Epidauros.*

v. Neumann, J. 1961. *The Mathematician.* Collected Works, Vol I, 1–9. Pergamon Press. (Originally 1947).

von Steuben, H. 1973. *Der Kanon des Polyklet.* Tübingen: Wasmuth.

Waschkies, H. J. 1971. Eine neue Hypothese zur Entdeckung der inkommensurablen Größen durch die Griechen. *Archive for History of Exact Sciences* 7, 325–353.

Waschkies, H. J. 1977. *Von Eudoxos zu Aristoteles.* Amsterdam: Grüner.

Waschkies, H. J. 1989. *Anfänge der Arithmetik im alten Orient und bei den Griechen.* Amsterdam: Grüner.

Waterhouse, W. C. 1972. "The discovery of the regular solids." *Archive for History of Exact Sciences* **9**, 212–221.

Wedberg, A. 1955. *Plato's philosophy of mathematics.* Stockholm: Almqvist & Wiksell.

Zeising, A. 1855. *Aesthetik.* Leipzig.

Zeuthen, H.G. 1915. *Sur l'origine historique de la connaissance des quantiteés irrationnelles.* Oversigt over det K. Danski Videnskabernes Forhandlinger, Copenhagen, 333–362.

Index